Peer-Led Team Learning

General, Organic and Biological Chemistry

The Workshop Project

Sponsored by the National Science Foundation

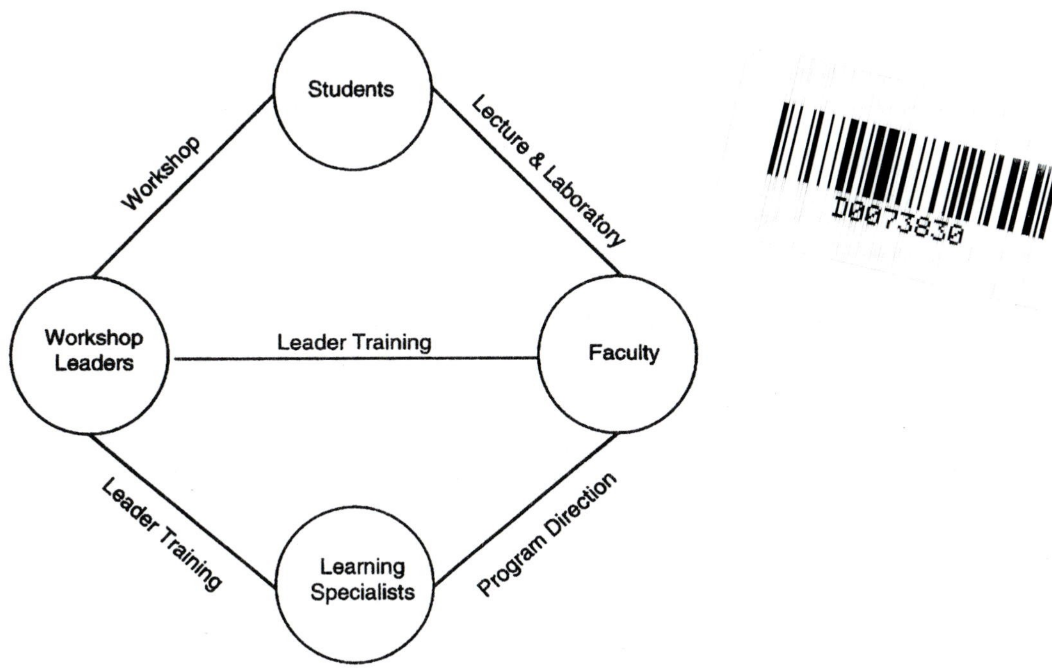

Pratibah Varma-Nelson • Mark S. Cracolice

Prentice Hall Series in Educational Innovation

PRENTICE HALL, Upper Saddle River, NJ 07458

Executive Editor: ***John Challice***
Project Manager: ***Kristen Kaiser***
Editorial Assistant: ***Gillian Buonanno***
Special Projects Manager: ***Barbara A. Murray***
Production Editor: ***Benjamin St. Jacques***
Manufacturing Manager: ***Trudy Pisciotti***
Manufacturing Buyer: ***Michael Bell***
Supplement Cover Manager: ***Jayne Conte***
Supplement Cover Designer: ***Maureen Eide***

Upper Saddle River, NJ 07458

Printed in the United States of America

10 9 8 7 6 5 4 3 2 1

ISBN 0-13-028361-4

Prentice-Hall International (UK) Limited, *London*
Prentice-Hall of Australia Pty. Limited, *Sydney*
Prentice-Hall Canada, Inc., *Toronto*
Prentice-Hall Hispanoamericana, S.A., *Mexico*
Prentice-Hall of India Private Limited, *New Delhi*
Prentice-Hall (Singapore) Pte. Ltd.
Prentice-Hall of Japan, Inc., *Tokyo*
Editora Prentice-Hall do Brasil, Ltda., *Rio de Janeiro*

Preface to the Peer-Led Team Learning Series

The Workshop Chemistry Project was an exploration, development, and application of the concept of peer-led team learning in problem-solving Workshops in introductory chemistry courses. A pilot project was first supported by the National Science Foundation, Division of Undergraduate Education, in 1991. In 1995, the Workshop Chemistry Project was selected by NSF/DUE as one of five systemic initiatives to "change the way introductory chemistry is taught." In the period 1991–1998, the project grew from the initial explorations at the City College of New York to a national activity involving more than 50 faculty members at a diverse group of more than 30 colleges and universities. In 1998–1999, approximately 2500 students were guided in Workshop courses by 300 peer leaders per term. In Fall 1999, NSF chose the Workshop Project for a National Dissemination Grant to substantially broaden the chemistry participation and to extend the model to other SMET disciplines, including biology, physics, and mathematics.

Peer-Led Team Learning—A Guidebook is the first of a series of five publications that report the work of the Project during the systemic initiative award (1995–1999). The purpose of these five books is to lower the energy barrier to new implementations of the model. The *Guidebook* is a comprehensive account that works back and forth from the conceptual and theoretical foundations of the model to reports of "best-practice" implementation and application. Three other books provide specific materials for use in Workshops in *General Chemistry*, *Organic Chemistry*, and *General, Organic, and Biological Chemistry*. One book in the series, *On Becoming a Peer Leader*, provides materials for leader training.

The collaboration of students, faculty, and learning specialists is a central feature of the Workshop model. The project has been enriched by the talents and energy of many participants. Some of their names are found throughout these books; many others are not identified. In either case, we are most grateful to all those who have advanced the model by their keen insight and enthusiastic commitment.

We also acknowledge, with pleasure, the support of the National Science Foundation, NSF/DUE 9450627 and NSF/DUE 9455920. Our work on the second NSF award was skillfully guided by our National Visiting Committee: Michael Gaines (chair), Joseph Casanova, Patricia Cuniff, David Evans, Eli Fromm, John Johnson, Bonnie Kaiser, Clark Landis, Kathleen Parson, Arlene Russel, Frank Sutman, Jeffrey Steinfeld, and Ronald Thornton. We value their advice and encouragement. The text of the *Guidebook* was repeatedly processed by Arlene Bristol, with exceptional skill and remarkable patience. Finally, we appreciate the vision and commitment of John Challice and Prentice Hall to make this work readily available to a large audience.

Books are written for you, the readers. We welcome your comments and insights. Please contact us at the indicated email addresses.

The Editors, Summer 2000

David K. Gosser	gosser@scisun.sci.ccny.cuny.edu
Mark S. Cracolice	markc@selway.umt.edu
J. A. Kampmeier	kamp@chem.rochester.edu
Vicki Roth	vrth@mail.rochester.edu
Victor S. Strozak	vstrozak@gc.cuny.edu
Pratibha Varma-Nelson	varmanelson@sxu.edu

Introduction to General, Organic and Biological Chemistry

Peer-led team learning is a unique approach to curriculum design in college science and mathematics. In this manual, we provide written materials to be used in a general, organic, and biological chemistry course. Most institutions teach this as a one-term or one-year survey course for nursing, wildlife biology, medical technology, environmental studies, nutrition, physical therapy, or other health science majors. We have attempted to provide more than the minimum number of units necessary for weekly meetings in a full-academic-year course. We believe that the units are representative of a standard course and are appropriate in the order presented, but you may wish to supplement our materials with your own, either by modifying the existing materials or by adding new material. We encourage you to pick and choose the units most appropriate for your course, using them in any order that is suitable. We have designed the units to the best of our ability to be flexible enough to fit into a wide variety of curriculum designs.

The units have been developed to be consistent with the philosophy of the project. The questions are designed to be answered by peer-led teams, not individuals. A group of six to eight students under the guidance of a peer leader should interact to develop solutions to the questions posed. Answers are not provided to keep the emphasis during the session on the method by which the question is answered rather than on the answer itself. We hope to encourage students to analyze, debate, and discuss the concepts underlying each question and also learn to decide the level of confidence they have when coming to a conclusion. In science, there are no absolute answers—only varying degrees of truth. This project is partly designed to convey the nature of scientific inquiry.

The level of confidence we have in the validity of the peer-led team learning approach increases with each new piece of evidence as we continue to assess the model. Our data strongly show that students learn more, have better attitudes toward learning, and have improved communication skills as a result of being actively engaged in the process of solving problems. We also believe that it is critical to build students' teamwork skills, as this is the model of the present and the future in the workplace. What better way to work on developing these teamwork skills than to have students work in teams?

Our deepest gratitude goes out to the students with whom we have worked during the development of these materials. The input of many of our students has led to numerous changes to improve the book. Two students in particular have made extensive contributions to the manuscript. Our thanks go to Dawn Patitucci and Jeff Trautmann for their valuable insight into learning and their knowledge of the subject. We also gratefully acknowledge the support of the National Science Foundation for their assistance in funding the project.

We welcome your comments, suggestions, and corrections. We also would appreciate input from your students and leaders. Please feel free to write directly on a copy of the pages and mail them to us. Comments should be sent to the senior author:

Pratibha Varma-Nelson
Science Department
Saint Xavier University
3700 West 103rd Street
Chicago, IL 60655

Voice: (773) 298-3526
Fax: (773) 779-9061
Email: varmanelson@sxu.edu

To the Instructor

We have sequenced the curriculum so that the peer-led team learning Workshop follows after the lecture and the initial student studying of the material in the textbook. We therefore expect students to be prepared with a basic knowledge of the concepts before coming to the Workshop. To encourage students to be prepared for their Workshop experience, we have found that administering a brief quiz at the beginning of the period helps students understand that they should have a fundamental knowledge of the concepts before participating in the Workshop.

Many years of project evaluation, led by Leo Gafney, have been distilled into six components that are critical to the success of the peer-led team learning model. Although we encourage you to read the evaluation chapter of the *Guidebook* (available from your Prentice Hall representative), we summarize these components here because of their importance.

- The organizational arrangements need to promote learning. These include the size of the group, quality of space, length of time, noise level, and availability of teaching resources.

- The Workshop materials are challenging at an appropriate level and are integrated with the other course components. They must be intended to encourage active learning, and they must work well in collaborative learning groups.

- The peer leaders are students who have successfully completed the course. They are well trained and closely supervised, with emphasis on knowledge of the Workshop problems, teaching/learning strategies, and leadership skills for small groups.

- The faculty teaching the courses are closely involved with the peer-led team learning Workshops and the peer leaders.

- The peer-led team learning Workshop sessions are integral to the course and coordinated with other elements of the course.

- The institution, at the highest levels of administration, and at the departmental level, encourages innovative teaching and provides sufficient logistical and financial support.

To the Workshop Leader

Congratulations on being selected for this important role! You will undergo training at your institution, which is of utmost importance in your development as a Workshop leader. We would like to add a few suggestions that have been formulated as a result of our experiences.

- Rehearse the Workshops with other leaders before meeting students, working through all the problems.

- Keep the focus of your team on how to get to answers rather than the answers themselves. Many students in introductory courses fixate on the answer, often without understanding the principles and concepts that are at the heart of the question. The approach to problem solving is of the utmost importance.

- Almost all chemistry problems are word problems. Be sure that students understand all the terminology and the question itself.

- Remind and encourage students to bring lecture notes, their textbooks, and their molecular model sets to the Workshops.

We have developed a number of tools to use in Workshops that promote group learning. We encourage you to use these as often as possible in your sessions.

Brainstorming is a problem-solving technique in which the group suggests ideas that can be used to solve a problem, initially without analysis. You can serve as the recorder as students give their ideas. After collecting a number of potential solutions, go back and analyze how they may be applied.

With the **round-robin** method, individual students contribute one step of a multistep solution, one at a time. This works well for long algebraic solutions, and it is also a good technique to use at the beginning of a term, when students are just starting to learn to work as a team.

Some problems are better suited for groups smaller than six to eight. We break the bigger group into groups of three or four or pairs on occasion. This works well with problems involving molecular models or organic synthesis, for example. Pairs can be designed so that a stronger student is paired with a weaker student. When you use **subgrouping**, it is important to have a representative from each subgroup discuss its results with the entire team to summarize and consolidate the concept.

The final question in each unit is designed to have students **reflect on their change in understanding**. These are particularly important. "Wrong" ideas often do not go away unless they are challenged. Students must realize that they have changed their understanding as a result of their Workshop experience to replace the older incorrect thinking with the new, improved comprehension. Try to reserve some time at the end of each Workshop for students to discuss how they have changed.

To the Student

This book supports a new way of learning with which you may have limited experience. We have found that some students are initially uncomfortable with peer-led team learning, mostly because they have worked by themselves throughout their entire academic experience. Give this method an open-minded chance, and you will probably find that it will help you learn chemistry at a deeper level than you can with traditional instructional techniques alone. You will also probably find that it is more enjoyable than always working by yourself. Our experience and research, and that of our colleagues, has shown that this method is very effective. Many students have benefited from this approach, and we hope that you will too.

After a number of years of listening to our students comment on peer-led team learning, we would like to pass along a few suggestions that they have felt would be beneficial to future students.

- Attend lecture. Workshops are not meant to replace lecture. New concepts are usually presented first by your instructor in lecture. Your team then will work together to gain a rich understanding of the concepts in Workshop.

- Bring your lecture notes, textbook, and molecular model kit to Workshop. You will frequently use these resources while solving Workshop problems.

- Prepare for Workshops, but do not start on the Workshop problems themselves until you are with your team. The *Topics to Review* section at the beginning of each unit is designed to help you focus on the things you should know before attending Workshop. It is important to prepare, as all members of your team depend on your contributions.

- Do not ask your Workshop leader for the "answer" to the questions. Answer keys do not exist! Your leader is a discussion facilitator, not an expert in chemistry. As you work through questions keep in mind that your goal is to learn how to answer questions of that type rather than to find the answer to the specific question. Work with your group to challenge your final conclusions, and do not stop until you can convince yourself and your team that your conceptual understanding is correct.

- Attend Workshops! Many years of evaluation show that this method works on many levels, including the near-term goal of improving your understanding of chemistry.

Table of Contents

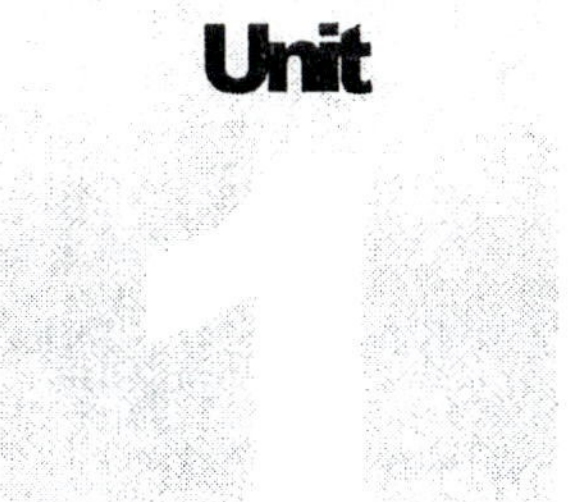

Measurements and Unit Conversions

Topics to Review

Instructions: Workshop chemistry works best if you come to Workshop prepared with a fundamental knowledge of the topics to be covered. Each Workshop chemistry unit therefore begins with a brief list of topics for you to review from your textbook and/or class notes. Before you attend Workshop, find the section in your textbook that addresses each of the review questions, spend a few minutes reviewing each topic until you have a basic understanding of the fundamental principles, and then write in the spaces provided the page numbers from your textbook where this topic is presented. Also write a brief answer to each question. Always bring your textbook to Workshop, and use this *Topics to Review* page as a quick reference to the appropriate section in your textbook during Workshop, as necessary.

1. What are the fundamental metric units used to measure mass (weight), length, and volume?

   Textbook pages _____ *to* _____.

2. What are the metric prefixes commonly used in chemistry?

   Textbook pages _____ *to* _____.

3. Briefly describe the problem-solving method given in your textbook for questions that involve conversions among units.

   Textbook pages _____ *to* _____.

4. Define density, both in words and in mathematical symbols.

   Textbook pages _____ *to* _____.

5. Define the terms energy (in general), potential energy, and kinetic energy.

Textbook pages _____ *to* _____.

6. Mathematically express the relationships among heat flow, mass, and change in temperature for a pure substance.

Textbook pages _____ *to* _____.

Questions

1. Answer each of the following questions and justify each answer.

   a. Is your mass (weight) best expressed in kilograms, grams, or milligrams?

   b. Is the volume of a typical drinking glass best expressed in kiloliters, liters, or milliliters?

   c. Is the length of a football field best expressed in kilometers, meters, or millimeters?

2. Rank each of the following from smallest to largest quantity:

a. 0.00445 kg, 3.88 g, 3905 mg

b. 35.5 mL, 0.0194 L, 8.5×10^{-5} kL

c. 833.1 cg, 9.2×10^{3} mg, 7.0×10^{-3} kg

3. Rank each of the following from smallest to largest quantity:

a. 11.5 in., 28.3 cm, 0.90 ft (1 in. ≡ 2.54 cm*)

b. 5.9 L, 6.11×10^3 cm^3, 1.75 gal (1 gal = 3.785 L)

c. 377 g, 2.577×10^8 μg, 0.91 lb (1 lb = 453.6 g)

* The triple bar, ≡, is used to identify a definition.

4. Use the dimensional analysis (unit conversion, factor label) problem-solving method to answer the following questions.

   a. How many nickels would you get for a twenty-dollar bill?

   b. How many hours are in a week?

   c. How many revolutions does the hour hand on a clock make in a year?

5. What is the density of a piece of wood that is 31.0 cm long, 5.2 cm wide, and 1.1 cm thick, with a mass of 163 g? Will it sink or float in water? Explain.

6. The density of water at 37°C is 0.993 g/mL. The density of normal urine ranges from 1.003 to 1.030 g/mL. Why is the density of urine greater than the density of pure water? An instrument known as a *hydrometer* is used to measure the density of urine samples in medical laboratories. What possible health problems could be indicated by abnormal urine densities?

7. A pure gold coin is dropped into a graduated cylinder, raising the level of the water in the cylinder from 10.00 mL to 11.24 mL. The density of gold is 18.88 g/mL. What is the mass of the coin?

8. A baseball is thrown straight up in the air and caught as it returns to the thrower's hand.

   a. At what point during the baseball's flight is its potential energy at a maximum? At a minimum? Explain.

   b. At what point during the baseball's flight is its kinetic energy at a maximum? At a minimum? Explain.

Questions 9–12: Consider the following list of specific heats:

Substance	Specific Heat (J/g · °C)
Aluminum (solid)	0.908
Ethanol (liquid)	2.45
Water (liquid)	4.18

9. Determine the quantity of heat that flows from a 25.0-g sheet of aluminum foil as it cools from an oven temperature of 450°F (232°C) to a room temperature of 70°F (21°C). Give the answer in kilojoules.

10. If all the heat flowing from the aluminum foil in Question 9 could be transferred to a 250-mL glass of water at room temperature, what final temperature would the water achieve?

11. What if the heat released from the aluminum in Question 9 was transferred to 250 mL of liquid ethanol at room temperature? What final temperature would it achieve? How does this value compare with the answer to Question 10?

12. Consider a human, who is composed of about 60% water, and a theoretical extraterrestrial humanoid who is similar to humans but has a body composition of 60% ethanol. Which being, the human or the ET, would have to expend less energy to maintain body temperature as the outside temperature changed from a day's high of 32°C to a night's low of 7°C? Explain your answer in terms of heat flows, and give mathematical examples to illustrate your solution to this question.

13. Reconsider your original answers in the *Topics to Review* section. Has your understanding of these topics changed as a result of this Workshop? Summarize your modified answers in the space below.

The Electronic Structure of Atoms

Topics to Review

Instructions for the *Topics to Review* sections are given at the beginning of Unit 1. If you are beginning your Workshops with Unit 2, please refer to the instructions in Unit 1.

1. What is the smallest particle of an element that retains the identity of that element?

   Textbook pages _____ *to* _____.

2. What is the smallest particle of a pure substance that retains the identity of that pure substance?

   Textbook pages _____ *to* _____.

3. What is an *element*?

Textbook pages _____ *to* _____.

4. What is a *compound*? How is it different from an element?

Textbook pages _____ *to* _____.

5. What are the charges, relative masses, and locations within the atom of the three subatomic particles?

Textbook pages _____ *to* _____.

6. What is the basis of the organization of elements in the periodic table?

Textbook pages _____ *to* _____.

7. What is an *orbital*?

Textbook pages _____ *to* _____.

8. What is an *electron configuration*?

Textbook pages _____ *to* _____.

Questions

1. What evidence exists to indicate that matter is made up of tiny particles? *Brainstorm* to come up with a list of types of evidence of the particulate nature of matter. Brainstorming is a problem-solving technique in which group members state their ideas freely, initially without regard to whether they are right or wrong. Assign a group member to be the recorder, and then spend a few minutes listing potential answers. Once you have compiled your initial list, go back to each idea and discuss its merits. Eliminate bad ideas and modify partially good ideas during the review and discussion. Once you have compiled an edited list, write a paragraph or two that would convince someone who has never taken a chemistry course that matter is *not* infinitely divisible.

2. Four words, each representing a concept, are given below. (a) Discuss the definition of each, one at a time, come up with a consensus definition, and then write that definition below each box. (b) Draw lines between related boxes, and write a description of the relationship on each line. For example, you might draw a line connecting the *atom* and *element* boxes, and write "the smallest particle of an element that retains the identity of that element is an atom." Draw as many connecting lines as possible, and come up with as many descriptions of the relationships among the concepts as possible.

Atom		Molecule

Element		Compound

3. If the weight of a proton were the same as the weight of a bowling ball (16 lb) and if the weights of the subatomic particles remained relatively the same, what would the weights of a neutron and electron be?

4. What percentage of the total weight of a carbon-12 atom (6 protons, 6 neutrons, 6 electrons) is found in the nucleus?

5. What process is used to organize the periodic table into rows?

What subatomic particle(s) is/are responsible for this sequence?

What process is used to organize the periodic table into columns?

What subatomic particle(s) is/are responsible for these groups?

6. Student A says that a good way to think about electron behavior is to imagine planets orbiting the sun. Student B disagrees, saying that orbitals must be considered when thinking about electrons. Divide your group in half, with each half arguing for one position. When you have come to a consensus, write a brief statement explaining which student is correct and why.

7. Sketch the shapes of a 1*s* and a 2*s* orbital. How are they similar? How are they different? What exactly do the orbitals represent? Do electrons follow paths along the orbital contours? Explain the relationship between the orbital shapes and the behavior of their associated electron(s). What are shells and subshells?

8. Your Workshop leader will assign one of the following pairs of elements to each student in your group:

a. Li and Na	b. Be and Mg	c. B and Al	d. C and Si
e. N and P	f. O and S	g. F and Cl	h. Ne and Ar

Begin by writing the complete ground state electron configuration for each of your elements. How are the electron configurations similar? How are they different? How will the electron configurations of other elements in the same periodic table group be similar and different? Can you find a relationship between the electron configurations and their group number? Discuss your conclusions with your workshop group.

9. Reconsider your original answers in the *Topics to Review* section. Has your understanding of these topics changed as a result of this Workshop? Summarize your modified answers in the space below.

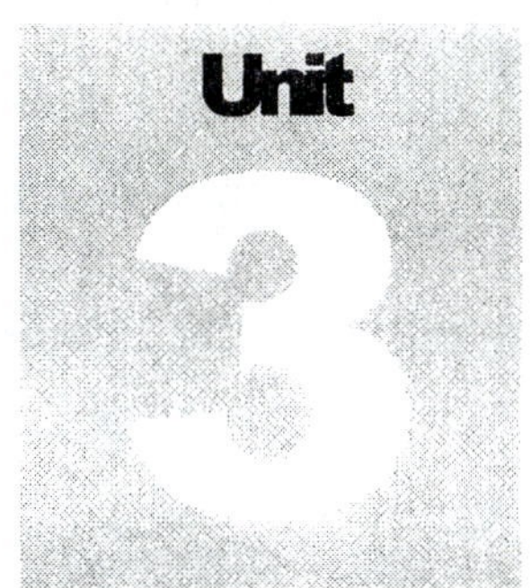

Nuclear Chemistry

Topics to Review

1. What is a radioactive material? How is it different from a nonradioactive material?

   Textbook pages _____ *to* _____.

2. Describe each of the following: alpha particles, beta particles, gamma radiation.

   Textbook pages _____ *to* _____.

3. Compare and contrast nuclear *fission* with nuclear *fusion*.

   Textbook pages _____ *to* _____.

4. What is the *half-life* of a radioactive substance?

   Textbook pages _____ *to* _____.

5. What units are used to express the rate of radioactive decay and the energy released and absorbed during such events?

   Textbook pages _____ *to* _____.

Questions

1. What is *radioactivity*? Are radioactive substances found on the surface of the earth or do they have to be created in a laboratory? Discuss the definition with your group and then come up with a paragraph that explains the concept to someone with no science background.

If you are given a rock and are told that it is radioactive, explain how you can test to see if it is. Assume that you are given access to equipment found at a major research university.

2. Choose one person from your group to serve as recorder, and set up two columns on the board, one with the heading "Ordinary Chemical Reactions" and one with the heading "Nuclear Reactions." Come up with at least three ways in which nuclear reactions are different from ordinary chemical reactions. List the differences side-by-side under the appropriate headings.

3. Complete the blanks in the following table with descriptions that are as complete as possible:

	Charge	Behavior in an Electrical Field	Mass (amu)	Nuclear Symbol	Penetrating Power	Composition in Terms of Subatomic Particles
Alpha Particles						
Beta Particles						
Gamma Rays						

4. Compare and contrast X-rays with alpha and beta particles, and with gamma rays.

5. What is *nuclear fission*? Carefully define this term and explain its importance to our society.

6. What is *nuclear fusion*? Carefully define this term and explain its importance to our society.

7. Compare how electrical power is generated in a conventional fossil-fuel-burning power plant with how it is produced in a nuclear power plant.

List the pros and cons of producing electricity by burning fossil fuels versus utilizing nuclear energy.

8. In this exercise, you will simulate the decay of a radioactive material in which 50% of the sample decays per day. You will work with a 100-atom sample of the radioactive material and use a coin flip to determine if any given atom decays on a particular day.

Divide the atoms equally among your group members (for example, if your group has seven members, each group is responsible for $100 \div 7 \cong 14$ atoms, and since $14 \times 7 = 98$, two groups will do 15 atoms). For each atom, flip a coin to determine if it decays on day 1. If the coin lands heads up, the atom does not decay, and if the coin lands tails up, the atom decays. Place an X in the day 1 box on the next page if the atom does not decay (heads). Leave the box blank if the atom undergoes radioactive decay (tails). Thus an undecayed atom will appear as a series of X's until it decays, after which it will have blank boxes.

When you have completed the exercise for ten days, total the number of atoms that have not decomposed on each day, and write the totals in the spaces provided.

Plot the number of undecayed atoms (*y*-axis) versus number of days (*x*-axis) on the following grid. This is essentially the same as a plot of fraction of sample remaining versus number of half-lives. Define half-life in terms of your plot and write your group's definition below your plot.

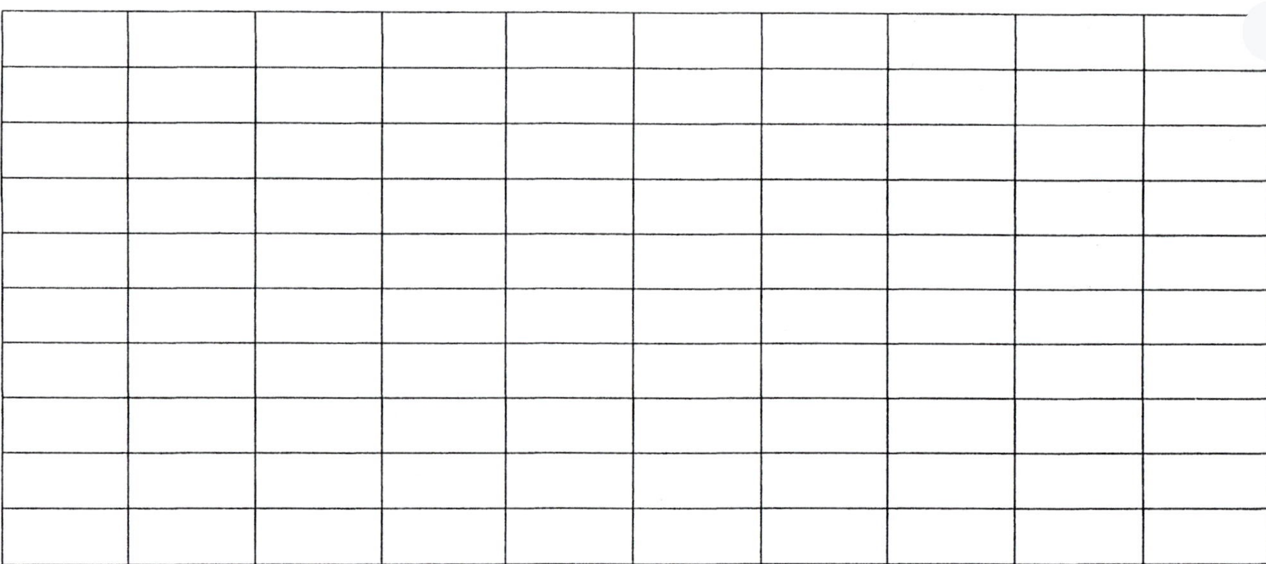

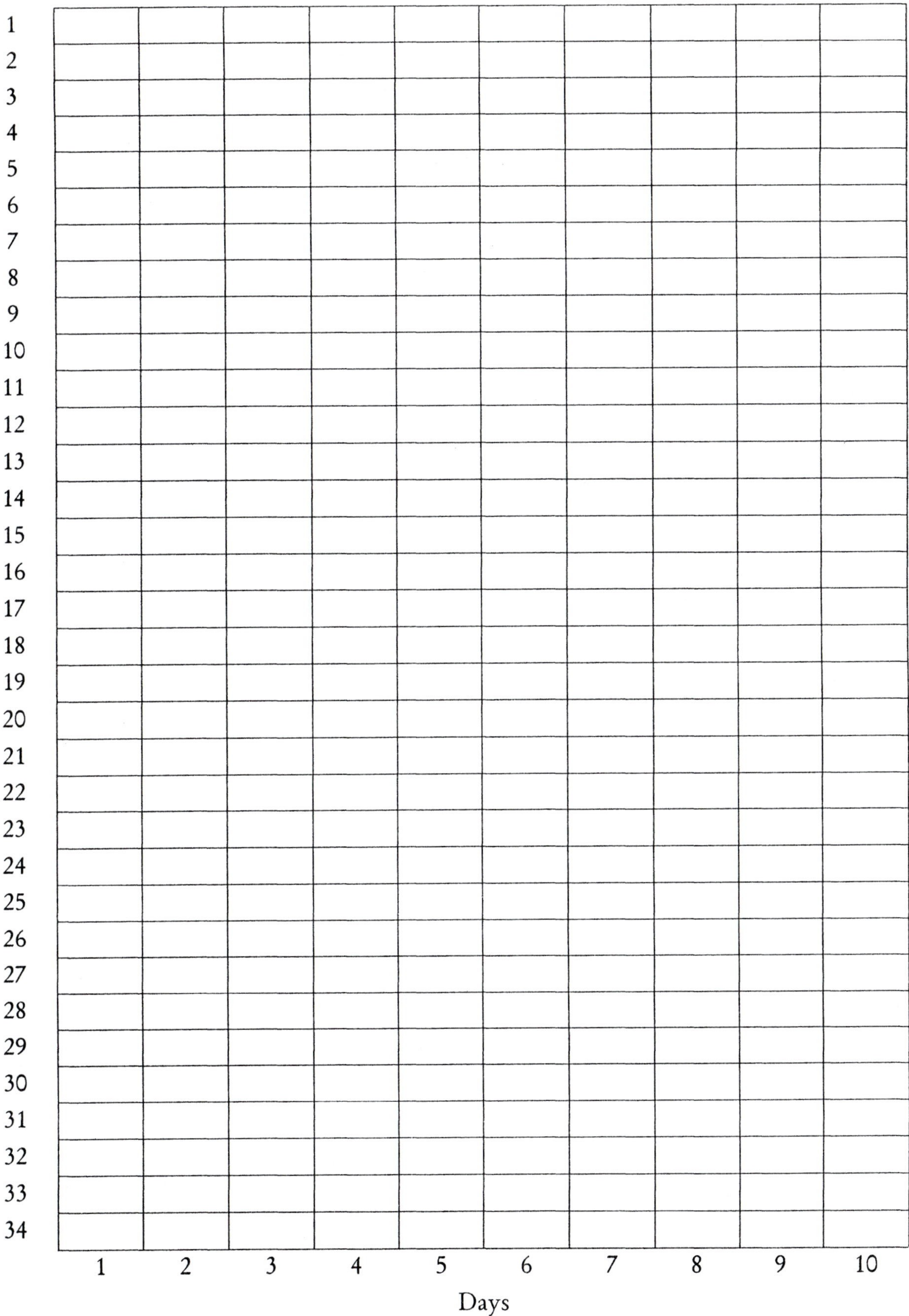

Atom	1	2	3	4	5	6	7	8	9	10
1										
2										
3										
4										
5										
6										
7										
8										
9										
10										
11										
12										
13										
14										
15										
16										
17										
18										
19										
20										
21										
22										
23										
24										
25										
26										
27										
28										
29										
30										
31										
32										
33										
34										

Days

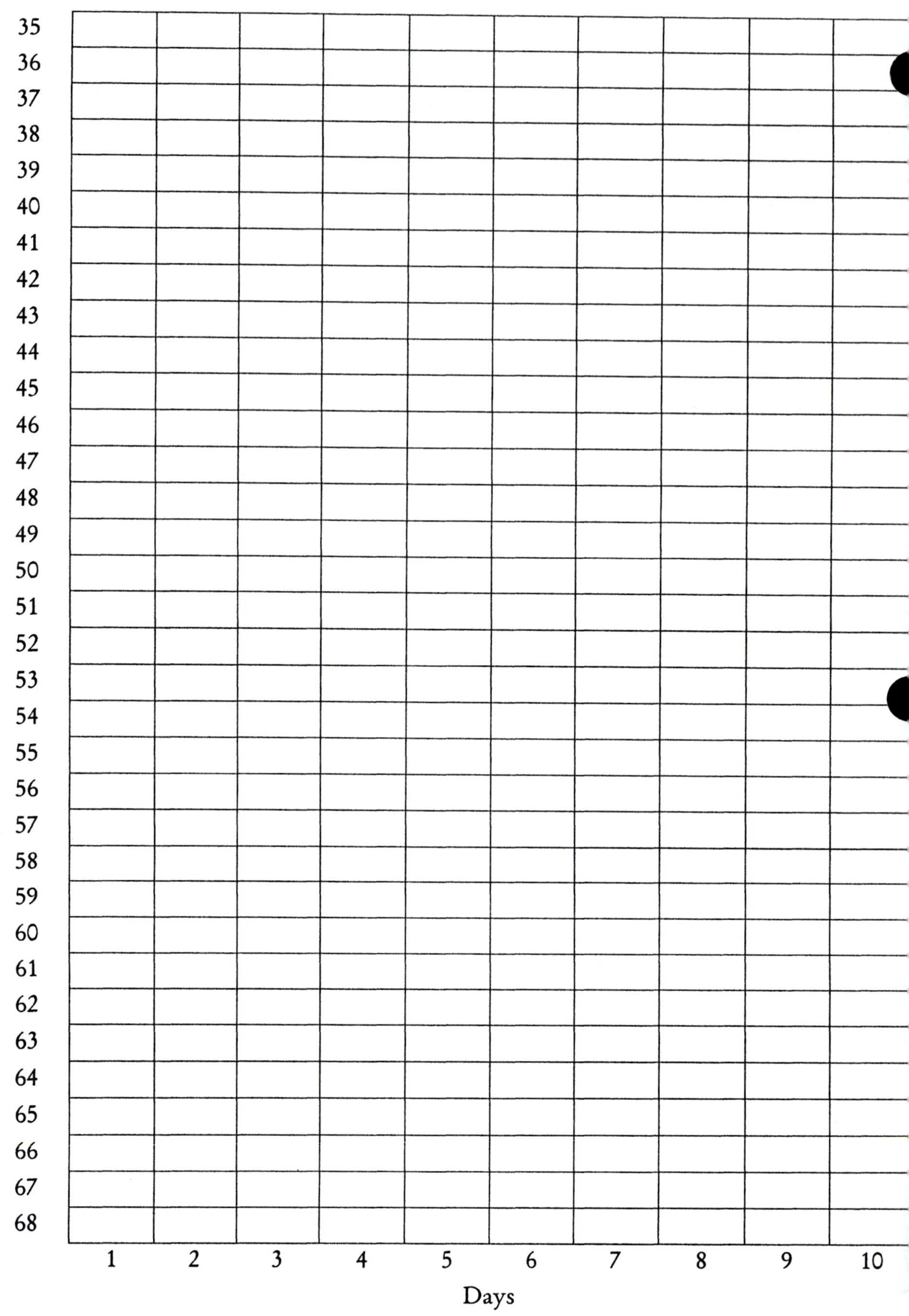
Atom
35
36
37
38
39
40
41
42
43
44
45
46
47
48
49
50
51
52
53
54
55
56
57
58
59
60
61
62
63
64
65
66
67
68
1
2
3
4
5
6
7
8
9
10
Days

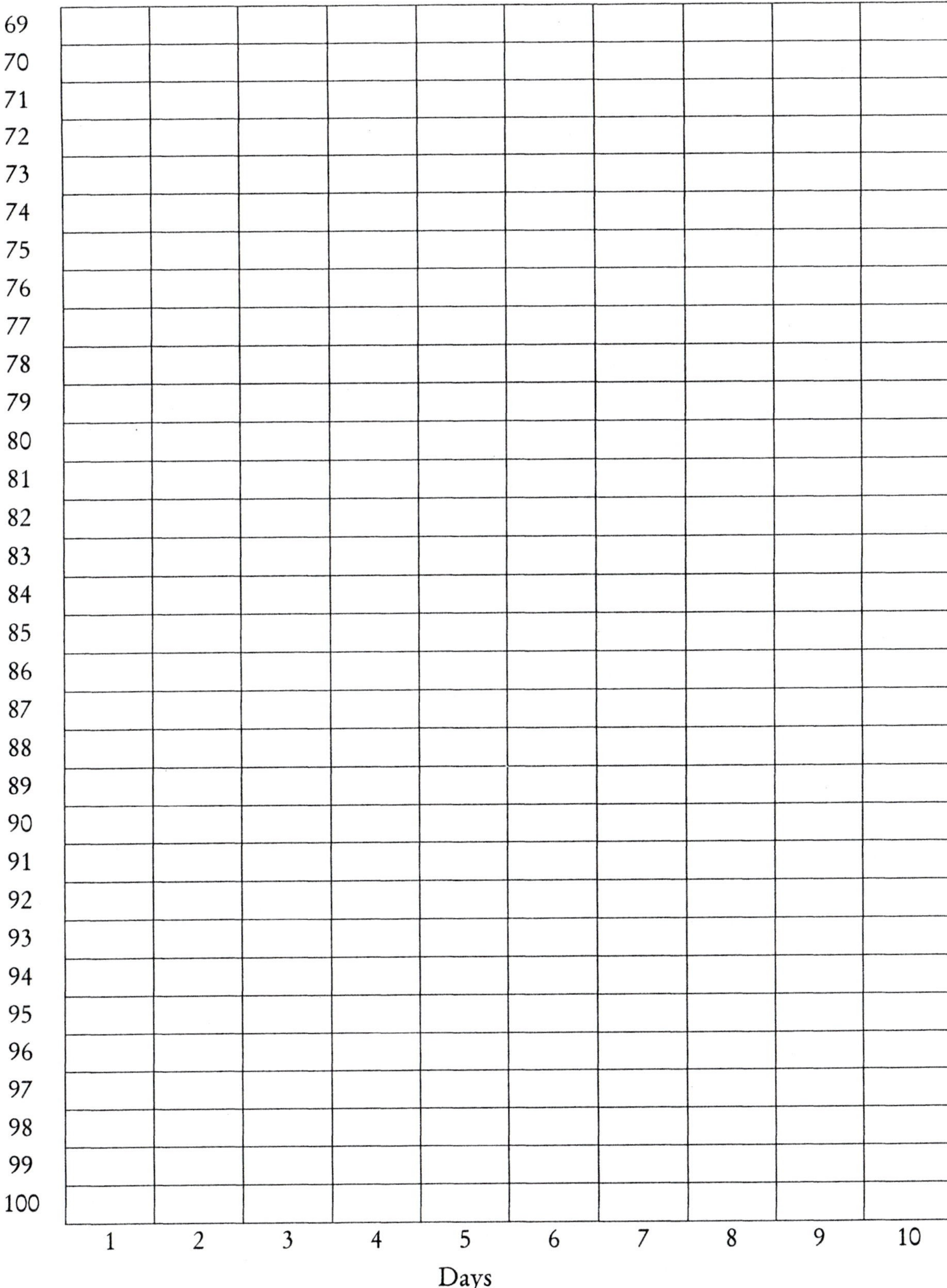

Total atoms undecayed _____ _____ _____ _____ _____ _____ _____ _____ _____ _____

9. Given that exposure to radiation is harmful to living organisms, why do people voluntarily expose themselves to such things as X-rays, smoke detectors, and nuclear medicine?

How can you protect yourself from exposure to background radiation?

10. The medical uses of radioactive substances can be categorized as either diagnostic or therapeutic. Diagnostic applications generally involve obtaining images of the distribution of radioactive compounds concentrated in a certain organ. For example, iodine-131 is commonly used for thyroid gland imaging. Therapeutic applications employ radioactive compounds to selectively destroy diseased cells. As an example, cobalt-60 is used to treat cancerous tumors. In general, diagnostic substances are gamma emitters, and therapeutic substances are beta emitters. Discuss and explain the relationship between the type of nuclear medical technique and the class of radioactive substance used.

11. Reconsider your original answers in the *Topics to Review* section. Has your understanding of these topics changed as a result of this Workshop? Summarize your modified answers in the space below.

Chemical Bonds

Topics to Review

1. What is an *ionic bond*?

   Textbook pages _____ to _____.

2. What is a *covalent bond*?

   Textbook pages _____ to _____.

3. What is a *polar* covalent bond? a *nonpolar* covalent bond?

   Textbook pages _____ *to* _____.

4. What is a *Lewis diagram*? You should be able to draw Lewis diagrams for ions and molecules formed from Period 1, 2, and 3 nonmetals *before* attending the workshop.

   Textbook pages _____ *to* _____.

5. Explain the consequences of valence shell electron pair repulsion (VSEPR) theory for determining molecular geometries from Lewis diagrams. What electron-pair geometry results from two regions of electron density around a central atom? three regions? four regions?

   Textbook pages _____ *to* _____.

Questions

1. Draw the Lewis symbol for a single atom of sodium. Draw the Lewis symbol for a single atom of chlorine. Use these symbols to illustrate the reaction of sodium and chlorine to form sodium chloride. Is the resulting bond ionic or covalent? Explain.

2. Draw the Lewis symbol for a single atom of fluorine. Draw the Lewis symbol for a single atom of bromine. Use these symbols to illustrate the reaction of fluorine and bromine to form bromine monofluoride. Is the resulting bond ionic or covalent? Explain.

3. Draw a particulate-level illustration of a potassium chloride crystal. Draw a particulate-level illustration of a calcium chloride crystal. How are the illustrations similar? How are they different?

4. Consider the formation of a hydrogen molecule beginning with two separated hydrogen atoms, $H + H \rightarrow H_2$. This process is known to proceed without outside intervention, and thus the hydrogen atoms must be naturally attracted to each other. Why? Consider the electrostatic forces involved to answer this question. Specifically, compare and contrast the attractive and repulsive electrostatic forces. Expand this discussion to the formation of covalent bonds in general.

5. Consider the following elements and their associated electronegativity values:

Li 1.0	Be 2.0	B 2.0	C 2.5	N 3.0	O 3.5	F 4.0
Na 0.9	Mg 1.2	Al 1.5	Si 1.8	P 2.1	S 2.5	Cl 3.0

Do electronegativity values follow a periodic trend?

What is the most polar compound that can be formed between two elements on this list? Does this compound have covalent or ionic bonding?

What is the least polar two-atom compound that can be formed? How would you classify the bonding in this compound?

Can you form a general rule regarding the position of elements in the periodic table and the type of bonding you will find between pairs of atoms of those elements?

6. Describe the difference between a polar covalent bond and a nonpolar covalent bond at the particulate level.

How can you predict whether a covalent bond will be polar or nonpolar? What role does electronegativity play in the answer to this question?

Give examples of a pair of atoms that share a polar covalent bond and a pair that share a nonpolar covalent bond.

What role does the symmetry of the bonding electron cloud have in bond polarity?

Your Workshop leader will assign a pair of nonmetal atoms to each member of the group. Determine if your atoms will form a polar or nonpolar bond and state whether the distribution of electronic charge will be symmetric or asymmetric. Explain your answers to the group.

7. Draw Lewis diagrams that show both the connections among the atoms and the lone-pair valence electrons for each of the following species. The molecular formulas give clues to the arrangements of the atoms in some cases.

CH_4

non-polar, tetrahedral

CH_3CH_2OH

polar, tetrahedral

H_2CCH_2

non-polar, trigonal planar

CH_3COOH

polar, trigonal planar

NO_3^-

Non polar, trigonal planar

H_2SO_4

VE: 2+8+24 = 32 electrons

8. Draw Lewis diagrams for each of the following: a chlorine atom, a chlorine molecule, and a chloride ion. Now consider the compound sodium chloride, which has ionic bonds, and hydrogen chloride, which has polar covalent bonds. What bonding type is found in the chlorine molecule?

Explain how chlorine can bond in all these different ways.

Discuss the differences among a chlorine atom, a chlorine molecule, and a chloride ion.

9. Discuss the relationship between the arrangement of electron pairs around a central atom in a molecule and the molecular geometry around that atom.

The electron pairs and bonds around the central atom determines the molecular geometry and bond angles in that atom.

How is molecular geometry affected by shared electron pairs versus lone electron pairs around a central atom?

How do single, double, and triple bonds affect molecular geometry around a central atom?

10. Predict the molecular geometry around each central atom for the Lewis diagrams from Question 7.

11. Reconsider your original answers in the *Topics to Review* section. Has your understanding of these topics changed as a result of this Workshop? Summarize your modified answers in the space below.

Stoichiometry: Formula Relationships

Topics to Review

1. How is the term *mole* defined in chemistry?

   Textbook pages _____ *to* _____.

2. What is Avogadro's number?

   Textbook pages _____ *to* _____.

3. What is the relationship between a mole and Avogadro's number?

Textbook pages _____ *to* _____.

4. How is *molar mass* defined?

Textbook pages _____ *to* _____.

5. How is molar mass calculated?

Textbook pages _____ *to* _____.

Questions

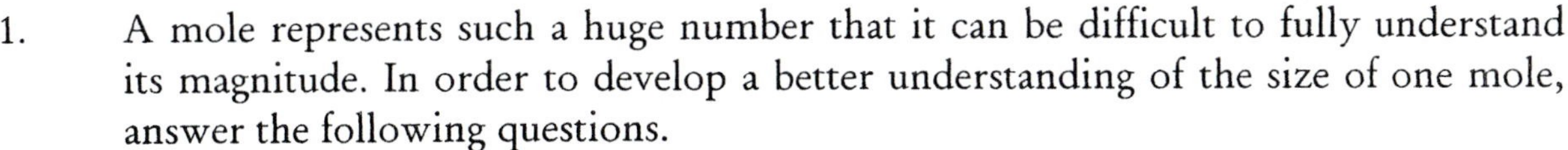

1. A mole represents such a huge number that it can be difficult to fully understand its magnitude. In order to develop a better understanding of the size of one mole, answer the following questions.

   If you were given one mole of dollars at birth and allowed to spend 1/80 of your inheritance per year over your projected lifetime of 80 years, what would be your annual allowance?

   How many dollars per day would you be allowed to spend?

   If the entire population of the earth, which is approximately six billion, were put to work counting the number of atoms in twelve grams of carbon-12, how long would it take to complete the task? Assume that each person would work 8 hours per day, 250 days per year, and that each person could count three atoms per second.

2. How many moles of ethanol, CH_3CH_2OH, are equal to 3.37×10^{25} ethanol molecules?

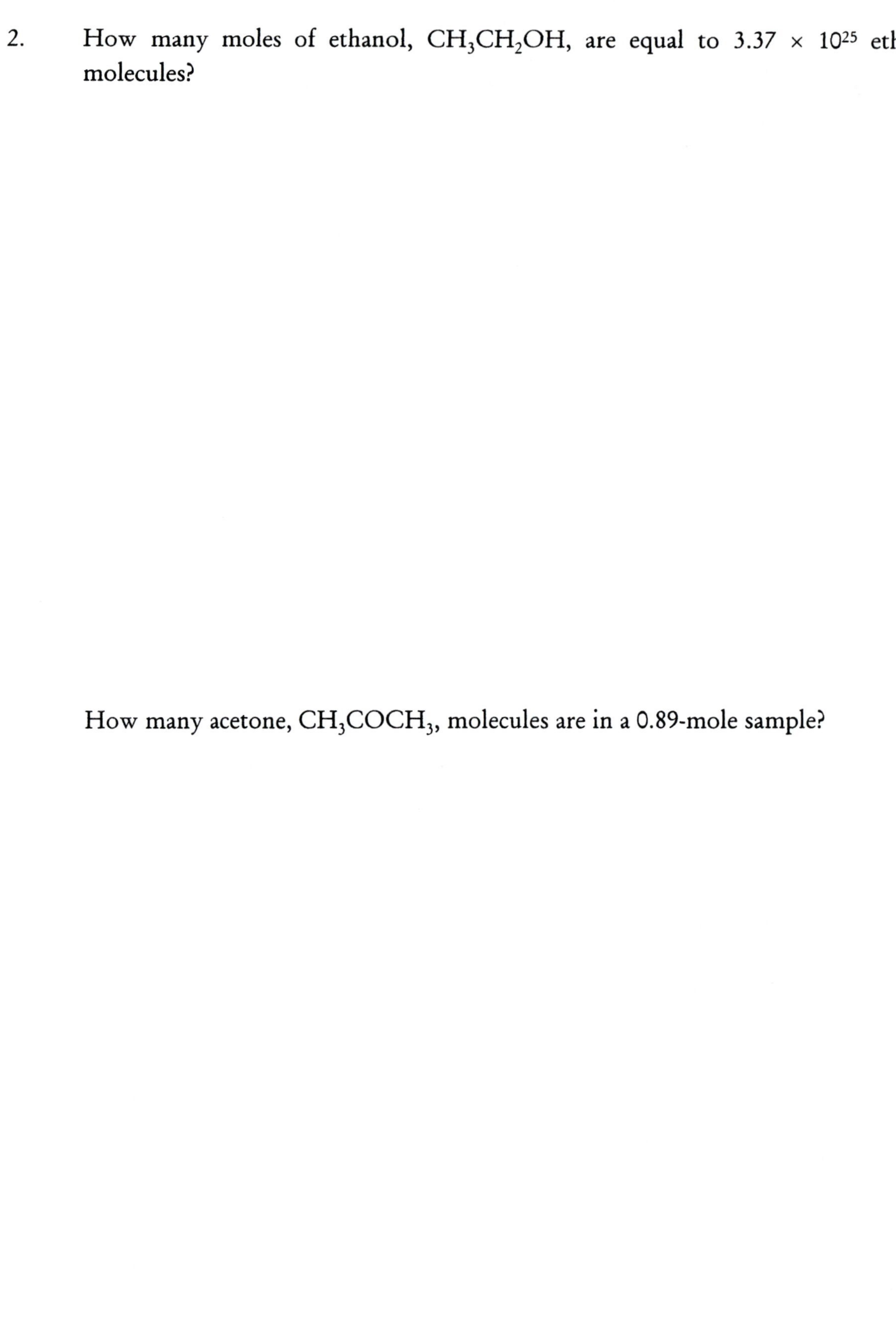

How many acetone, CH_3COCH_3, molecules are in a 0.89-mole sample?

3. Determine the molar mass of each of the following compounds:

Acetic acid, CH_3COOH (present in vinegar)

Glucose, $C_6H_{12}O_6$ (blood sugar)

Dimethyl ether, CH_3OCH_3 (anesthetic)

Butane, C_4H_{10} (lighter fluid)

Citric acid, $H_3C_6O_7H_5$ (essential for respiration)

4. What is the mass of 100 billion billion ($100 \times 10^9 \times 10^9$) water molecules?

5. The density of aniline, $C_6H_5NH_2$, is 1.02 g/mL. How many aniline molecules are in a 1.00-L sample?

6. Reconsider your original answers in the *Topics to Review* section. Has your understanding of these topics changed as a result of this Workshop? Summarize your modified answers in the space below.

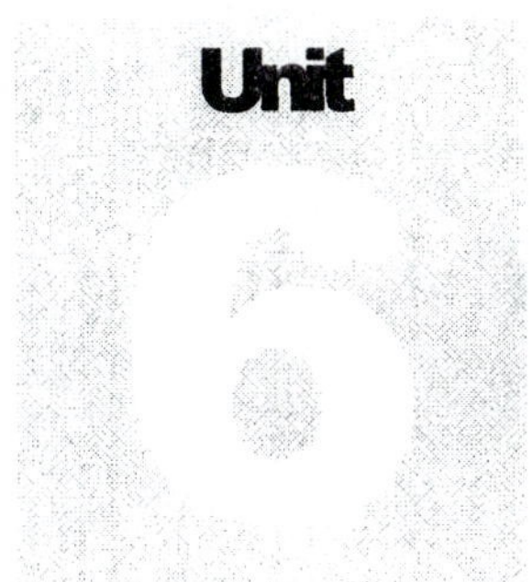

Stoichiometry: Chemical Reactions

Topics to Review

1. What is the relationship between the coefficients in a balanced chemical equation and the relative number of moles of each reactant and product in the reaction?

   Textbook pages _____ to _____.

2. Briefly describe the procedure by which you can calculate the mass of one species in a chemical change when given the mass of another species and a balanced chemical equation.

   Textbook pages _____ to _____.

3. How is percentage yield determined for a chemical reaction?

Textbook pages _____ to _____.

Questions

1. Consider the combustion of ethylene:

$$C_2H_4 + 3\,O_2 \rightarrow 2\,CO_2 + 2\,H_2O$$

How many moles of oxygen are needed to react with 3.5 moles of ethylene?

How many moles of ethylene reacted if 0.50 mole of water is produced?

How many moles of carbon dioxide will be produced when 2.8 moles of oxygen react?

2. When water is added to solid calcium carbide, CaC_2, acetylene gas, C_2H_2, and calcium hydroxide, $Ca(OH)_2$, are produced.

   Write a balanced equation for the reaction.

   How many moles of acetylene are produced when 0.50 mole of calcium carbide reacts?

   How many moles of calcium carbide are needed to completely react with 5.00×10^{22} water molecules?

3. Aspirin, $C_9H_8O_4$, can be made from the reaction of salicylic acid, $C_7H_6O_3$, and acetic anhydride, $C_4H_6O_3$. An additional product of the reaction is acetic acid, $C_2H_4O_2$. What is the minimum amount of salicylic acid necessary to produce 100 tablets which each contain 325 mg of aspirin?

4. Plants produce glucose through photosynthesis, which is the chemical reaction of carbon dioxide, CO_2, and water to yield glucose, $C_6H_{12}O_6$, and oxygen, O_2. Sunlight is the source of energy that drives the reaction. What is the minimum mass of carbon dioxide needed to produce 50.0 mg of glucose through a photosynthesis reaction? How much water is also required?

5. Under what conditions can a percentage yield be greater than 100%?

The maximum product yield for a large-scale industrial reaction is known to be 375 kg. What is the percentage yield if 315 kg is actually produced?

6. Plant fertilizers are made from ammonia, NH_3, which is made by the reaction of nitrogen, N_2, and hydrogen, H_2. If an industrial ammonia manufacturing plant can produce ammonia in a 84.7% yield, based on the mass of hydrogen reacting, what mass of hydrogen is required to produce 500.0 kg of ammonia?

7. Reconsider your original answers in the *Topics to Review* section. Has your understanding of these topics changed as a result of this Workshop? Summarize your modified answers in the space below.

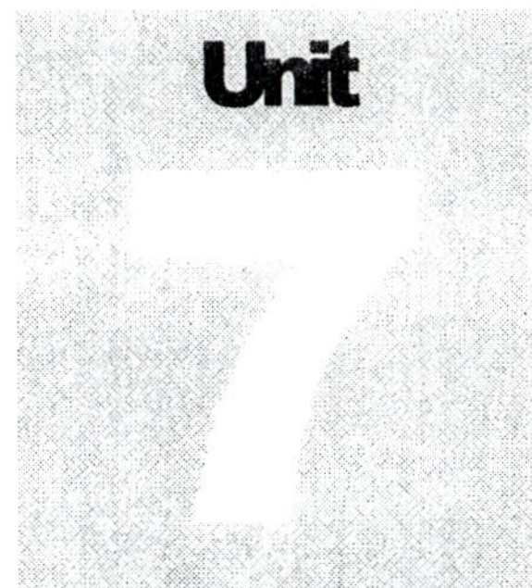

Gases

Topics to Review

1. Define *pressure*. What is the relationship between the pressure of a gas, a macroscopic property, and the behavior of the gas *particles*? In other words, how is the pressure of the entire gas sample related to the behavior of the particles that make up the sample?

   Textbook pages _____ *to* _____.

2. What pressure units are given in your textbook? What is the relationship among those units?

   Textbook pages _____ *to* _____.

3. What is the relationship between the Celsius temperature scale and the kelvin temperature scale?

Textbook pages _____ to _____.

4. What is *standard temperature and pressure*?

Textbook pages _____ to _____.

5. What is the algebraic relationship among the pressure, volume, and temperature of a fixed quantity of a gas?

Textbook pages _____ *to* _____.

6. What is the algebraic relationship among the pressure, volume, amount, and temperature of a gas?

Textbook pages _____ *to* _____.

Questions

1. Compare a barometer and a manometer. How does each work? Draw a sketch of each instrument, and use your sketches to explain how each instrument measures pressure. For what purpose is each used?

2. Consider an ice cube in an otherwise empty rectangular box. Draw a three-dimensional sketch of the ice cube in the box, illustrating the ice cube at the particulate level. In order to keep your illustration relatively simple, use spheres to represent water molecules.

Draw a new sketch showing the contents of the box after the ice melts to become liquid water.

Draw another sketch showing the contents of the box after the water is vigorously heated, turning it completely to steam.

What if the process was reversed? How would your sketches be different?

3. Discuss the behavior of the particles in the gaseous state, specifically answering the following questions:

Discuss how the particles move. Do they follow straight paths, curved paths, or both? Do they all move at the same speed? What is the relationship between temperature and speed of the particles?

Discuss what happens when particles collide. Do they stick together, "bounce" off one another, or both? What happens when they strike the container walls?

Discuss the relative size of the particles and the container size. How does the volume of empty space between particles compare with the volume of the particles themselves?

4. What is an ideal gas? How is it different from a real gas? How is it similar? Under what conditions is a real gas closely approximated by an ideal gas?

5. What gases make up the earth's atmosphere?

6. When a glass of ice water sits on a table, water often appears on the outside of the glass. Where does this water come from?

7. Convert each of the following pressures to atmospheres.

675 torr

723 mm Hg

28.7 inches of mercury

8. Convert each of the following temperatures to kelvins.

−254°C

35°C

72°F

9. Consider a flexible-walled rubber balloon that has been inflated and tied closed.

What happens to the balloon volume as it is heated? What happens to the volume as it is cooled? Express this relationship as an algebraic equation.

What happens to the pressure of the air in the balloon when its volume is reduced by squeezing it? What happens to the pressure when the volume is allowed to increase back to its original quantity? Express this relationship as an algebraic equation.

10. A flexible-walled container holds 1.75 L of helium gas.

a. What happens to the container volume when the absolute temperature is increased by a factor of 1.5 while the pressure remains constant?

b. What is the new volume if the pressure is doubled while the temperature is kept constant?

c. How is the volume affected when both of the changes given in parts (a) and (b) are made simultaneously?

11. A gas has a volume of 365 mL at 23°C and 717 mm Hg. What volume will it occupy at 36°C and 755 mm Hg?

12. Assume that air is 80% N_2 and 20% O_2. What is the average molar mass of air?

What mass of air is present in a person's lungs if the total lung volume is 2 L and the air is at 1 atm and 37°C?

13. Your Workshop leader will assign one of the following questions to you and a partner. Find the solution to your question *and* develop a generalized strategy to solve similar questions. When all pairs have solved their questions and developed strategies, discuss the solutions and strategies as a group, and then find a single general strategy to answer questions similar to these.

What is the pressure of a 0.85-L sample of He with a mass of 0.070 g when it is at 5°C and 1.2 atm?

A 5.0-g sample of N_2 occupies a flexible-walled container at 19°C and 705 torr. What is the volume of the container?

What mass of methane (CH_4) will be found in a cylinder with a 125-L capacity if the gas is at 3.7 atm and 15°C?

Determine the Celsius temperature of the contents of a 2.0-L oxygen (O_2) cylinder if the mass of the cylinder is 975 g before it is filled and 980 g after it is filled to a pressure of 1500 mm Hg.

14. a. Use the ideal gas law to determine the molar volume of an ideal gas at STP.

b. Use the molar volume from part (a) to calculate the volume of 2.0 mol of an ideal gas at STP.

c. How many moles of neon are in a 0.25-L container at STP?

15. Reconsider your original answers in the *Topics to Review* section. Has your understanding of these topics changed as a result of this Workshop? Summarize your modified answers in the space below.

Liquids and Solids

Topics to Review

1. Compare and contrast the distances between particles in the solid, liquid, and gaseous states. How do these distances affect intermolecular attractive forces?

   Textbook pages _____ *to* _____.

2. List the three major intermolecular forces in order of increasing strength.

   Textbook pages _____ *to* _____.

3. How can you predict the significant intermolecular forces that will affect the physical properties of a compound?

   Textbook pages _____ *to* _____.

4. The strength of intermolecular forces allows you to predict trends in *vapor pressure* and *boiling point* among otherwise similar compounds. Define each of these physical properties.

   Textbook pages _____ *to* _____.

5. Distinguish between a crystalline solid and an amorphous solid.

Textbook pages _____ *to* _____.

6. Define and compare heat (enthalpy) of vaporization and heat (enthalpy) of fusion.

Textbook pages _____ *to* _____.

Questions

1. What are the similarities and differences between *inter*molecular forces and *intra*molecular forces (chemical bonds)? Brainstorm to come up with a list of items that can be used to compare the forces, and then discuss the items on your group's list. (If you were not assigned Unit 2, see Question 1 in that unit for a description of the *brainstorming* problem-solving technique.) Finalize the answer to the question by listing at least two similarities and differences among the types of forces.

2. The term *hydrogen bonding* implies that this phenomenon is a chemical bond, yet it is listed in the "intermolecular forces" section of most textbooks. Is a hydrogen bond a bond or not? Explain.

What characteristics must be present in a molecule in order for hydrogen bonding to occur? Give three examples of molecules that will exhibit hydrogen bonding and three examples of molecules that will *not* exhibit hydrogen bonding *even though they have hydrogen atoms.*

3. The boiling points of the hydrogen compounds of the Group 6A (16) elements, from bottom to top, are:

H_2Te –4°C
H_2Se –42°C
H_2S –62°C

Extending this trend to H_2O, we would expect the boiling point of water to be about –72°C. Instead, it is +100°C. What makes water molecules different from the other Group 6A (16) hydrogen compounds? Explain. What is the significance of this phenomenon to living organisms?

4. Discuss the similarities between a polar molecule and a bar magnet. Summarize your discussion by writing a paragraph describing the particulate-level behavior of polar molecules.

What characteristics must be present in a molecule for it to be polar? Give three examples of polar molecules and three examples of nonpolar molecules.

5. Consider the following pairs of molecules, and explain why their boiling points are different.

SiH_4 (molecular mass = 32 amu), boiling point = −112°C and
PH_3 (molecular mass = 34 amu), boiling point = −85°C

GeH_4 (molecular mass = 77 amu), boiling point = −90°C and
AsH_3 (molecular mass = 78 amu), boiling point = −55°C

Br_2 (molecular mass = 160 amu), boiling point = 59°C and
ICl (molecular mass = 162 amu), boiling point = 97°C

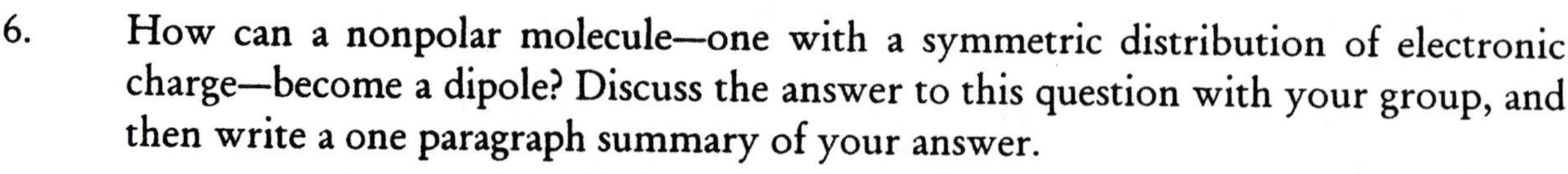

6. How can a nonpolar molecule—one with a symmetric distribution of electronic charge—become a dipole? Discuss the answer to this question with your group, and then write a one paragraph summary of your answer.

How do you identify nonpolar molecules? Can a molecule made from different elements, such as CH_4, be nonpolar? Are all molecules made from the same element, such as H_2, nonpolar? Explain.

7. C_2H_6, C_4H_{10}, and C_6H_{14} have similar molecular structures, with the carbon atoms arranged in a continuous chain. Arrange these compounds in order of increasing boiling point, and explain the reasoning behind your answer.

8. Define *boiling point.* To do so, begin by discussing this concept in pairs, and then discuss your understanding with your Workshop group. While working in pairs, compare your current understanding with the definition you would have given before taking a chemistry course. When you have completed your group discussion, write a paragraph explaining the boiling concept so that it can be understood by someone with no chemistry background.

9. What is vapor pressure? Draw a particulate-level illustration as part of your definition.

10. Compare and contrast crystalline and amorphous solids in terms of their particulate and macroscopic characteristics. Give examples of each.

11. If a bag of 100 marbles is emptied into a shoe box, will the arrangement most resemble a crystalline or an amorphous solid at the particulate level? Explain. How could you rearrange the marbles to have the opposite arrangement?

12. A glass is partially filled with ice and then liquid water is added. The contents are stirred occasionally over a period of ten minutes, after which both ice and liquid water are present in the glass. What is the temperature of the ice water? How do you know the answer to this question?

What will the temperature be after another ice cube is added to the mixture?

13. Consider the following plot of temperature versus energy for a pure substance.

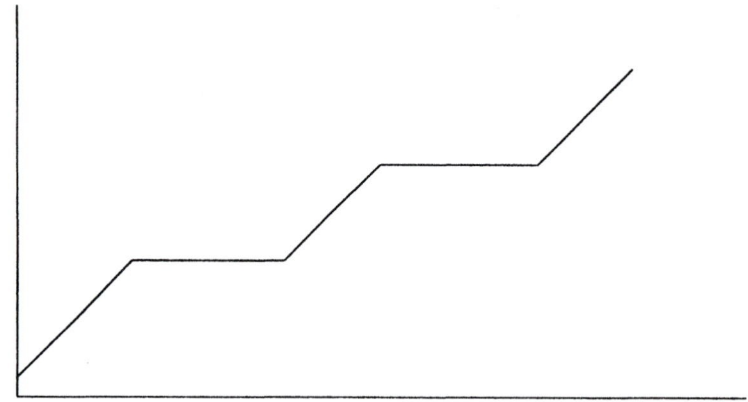

Which variable is plotted on which axis?

What state(s) of matter exist(s) in each of the five sections of the curve?

Explain why the two horizontal sections of the curve exist, in terms of particulate behavior.

14. 2.26 kJ of energy is released when 1.00 g of steam condenses to liquid water at 100°C. How much energy is released when 100.0 g of steam condenses to the liquid state at 100°C?

4.18 J of energy is released when 1.00 g of liquid water cools 1°C. How much energy is released when 100.0 g of liquid water at 100°C cools to 0°C?

335 J of energy is released when 1.00 g of liquid water freezes to become ice at 0°C. How much energy is released when 100.0 g of water freezes to the solid state at 0°C?

How much energy is released when 100.0 g of steam at 100°C is cooled to ice at 0°C?

15. Use the data from Question 14 as necessary to determine the quantity of energy required to change 1.0 kg of liquid water at 20°C to steam at 100°C.

16. Reconsider your original answers in the *Topics to Review* section. Has your understanding of these topics changed as a result of this Workshop? Summarize your modified answers in the space below.

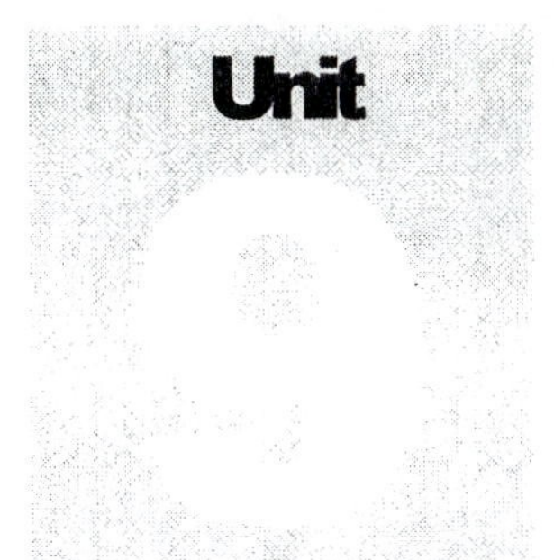

Solutions

Topics to Review

1. How do chemists define the term *solution*?

   Textbook pages _____ to _____.

2. Define the following terms as they apply to solution chemistry: *solute, solvent, miscible, immiscible, concentrated, dilute, supersaturated, saturated, unsaturated, soluble.*

   Textbook pages _____ to _____.

3. Explain how the solubility of any pair of substances can be predicted.

   Textbook pages _____ *to* _____.

4. Write the defining equations for each of the following: weight/weight percentage concentration, volume/volume percentage concentration, weight/volume percentage concentration, parts per million, parts per billion, molarity.

   Textbook pages _____ *to* _____.

5. What distinguishes colligative properties of solutions from those that are not colligative?

Textbook pages _____ *to* _____.

6. Define the term *osmotic pressure* and describe how it can be measured.

Textbook pages _____ *to* _____.

Questions

1. Give a precise chemical definition of the term *solution*. Also define any chemical terms that you use in your definition.

   a. Give at least two examples of solutions in each of the common states of matter: solid, liquid, and gas.

   b. Draw a particulate-level sketch of a salt (sodium chloride)-dissolved-in-water solution. What is the relationship between the definition of a solution and its particulate-level composition?

2. Define, compare, and contrast terms in each of the following groups.

   a. Solute and solvent

   b. Miscible and immiscible

   c. Concentrated and dilute

   d. Supersaturated, saturated, unsaturated, and solubility

3. Brainstorm to come up with a list of methods you could use to speed up the process of dissolving sugar in water. When you have completed a list, choose the three best methods from your list and give a particulate-level chemical explanation of why each method speeds up the dissolving process.

4. a. What is the meaning of each of the following terms?

Hypotonic solution

Hypertonic solution

Hemolysis

Crenation

b. Write a single paragraph using all the terms.

5. Which of the following are more soluble in water than in benzene (a liquid nonpolar solvent)? Which are more soluble in benzene than in water? (*Hint*: Draw Lewis structures of compounds for which a structure can be drawn.)

   a. Baking soda, $NaHCO_3(s)$

   b. Sulfur, $S_8(s)$

   c. Carbon tetrachloride, $CCl_4(\ell)$

   d. Ethanol, $CH_3CH_2OH(\ell)$

   e. *n*-Hexane, $C_6H_{14}(\ell)$

   f. Ammonia, $NH_3(g)$

6. The solubility of magnesium hydroxide in water is 0.0009 g/mL at 18°C. What is the minimum amount of water necessary to dissolve 1.0 g $Mg(OH)_2$? What mass of magnesium hydroxide will dissolve in 1.0 L of water?

7. Your group works in a chemical research laboratory. The following volumetric flasks are available: 10 mL, 25mL, 50 mL, 100 mL, 200 mL, 250 mL, 500 mL, 1000 mL, and 2000 mL. All flasks are accurate to ± 0.1 mL. You also have access to all the compounds needed, a source of distilled, deionized water, and a balance accurate to ± 1 mg. Carefully and precisely explain how you would make each of the following solutions:

   1.00 L of a 0.70 M solution of sodium chloride (table salt), NaCl

   0.50 L of a 2.5% w/w solution of sucrose (table sugar), $C_{12}H_{22}O_{11}$

   0.10 L of a 10.0% v/v solution of ethanol, C_2H_5OH

   0.40 L of a 5.0% w/v solution of ammonium phosphate, $(NH_4)_3PO_4$

8. Determine the molarity of a solution that is 88.2 ppm in urea, $CO(NH_2)_2$.

9. Each mole of particles added to one kilogram of water lowers the freezing point by 1.9°C. Determine the freezing point of solutions made by adding the following quantities to one kilogram of water:

a. 
16.0 g methanol, CH_3OH

b. 58.44 g sodium chloride, NaCl

c. 166.5 g calcium chloride, $CaCl_2$

If each of the preceding substances had the same cost per kilogram, which one would be the best choice for melting ice in the winter? Explain.

10. Define the following terms: *semipermeable membrane, osmotic pressure.*

a. Give a real-life example of each. (*Hint*: Check your textbook for examples if you cannot think of any from your everyday experiences.)

b. What is the difference between an osmotic membrane and a dialysis membrane?

c. Why is 5.5% glucose solution used for intravenous feeding? Why is a 0.89% sodium chloride solution used for intravenous injections? How were these quantities determined? Why are they not the same? What are the consequences of using other concentrations?

11. Reconsider your original answers in the *Topics to Review* section. Has your understanding of these topics changed as a result of this Workshop? Summarize your modified answers in the space below.

Chemical Kinetics & Equilibrium

Topics to Review

1. Define the term *chemical kinetics.*

   Textbook pages _____ *to* _____.

2. What conditions must be satisfied for a reaction-producing molecular collision to occur?

   Textbook pages _____ *to* _____.

3. What is *activation energy*? What is its role in affecting the rate of a chemical reaction?

   Textbook pages _____ *to* _____.

4. What factors can be controlled to help reactants over an activation energy barrier and thus increase the reaction rate?

   Textbook pages _____ *to* _____.

5. Why do chemical reactions proceed faster at higher temperatures?

   Textbook pages _____ *to* _____.

6. Explain what a catalyst is and how it works.

Textbook pages _____ *to* _____.

7. What is equal in a chemical equilibrium?

Textbook pages _____ *to* _____.

8. Write the mathematical definition of the equilibrium constant, K.

Textbook pages _____ *to* _____.

9. Write a brief summary of Le Chatelier's principle.

Textbook pages _____ *to* _____.

Questions

1. Consider the multitude of chemical reactions that occur in everyday life (exclude those that usually occur only in a chemistry laboratory) and the speed at which those reactions occur.

   a. At the qualitative level only, describe at least two reactions that occur relatively slowly and two reactions that occur relatively quickly.

   b. What term do chemists use to describe the study of the speed at which reactions occur?

   c. In what units are rates of change for chemical compounds expressed? Give a general description of the units and some specific examples. (For example, if this question was asking about rates of automobile speed, they are generally expressed in distance units per time units; examples are miles per hour and meters per second.)

2. At the particulate level, the effectiveness of molecular collisions is responsible for determining the rate of chemical reactions. Whether or not a collision is effective is in turn primarily dependent on the energy of the colliding molecules and their orientation.

   a. Why is a large amount of collision energy between particles necessary for a chemical reaction to occur?

   b. Consider the reaction $H_2(g) + F_2(g) \rightarrow 2\ HF(g)$. Sketch the reacting molecules, showing some orientations that will result in a successful reaction and some that will not. If orientations of colliding molecules are random, what percentage of collisions would you guess would have the proper orientation for this reaction?

3. Sketch two plots of energy (*y*-axis) versus reaction progress (*x*-axis), one for an endothermic reaction and one for an exothermic reaction. On each diagram, label the following: energy of reactants, energy of products, energy of reaction, activation energy, transition state, and activated complex.

4. Four factors are primarily responsible for the rate of a chemical reaction: the nature of the reactant molecules, their concentration, the temperature, and the presence of a catalyst. Explain how each affects the reaction rate.

5. a. Draw a plot of fraction of molecules with a given kinetic energy (*y*-axis) versus kinetic energy (*x*-axis) for a typical compound. Describe what is represented by the area under the curve.

   b. Indicate on the *x*-axis a typical value for the minimum energy required for a successful collision. Discuss the meaning of the area under the curve and to the right of this value.

   c. Draw a curve representing the kinetic energy distribution of the same compound at a higher temperature on the same axes. Use the plot to explain why reaction rates increase with increasing temperature.

6. Sketch a plot of energy (y-axis) versus reaction progress (x-axis) for a typical exothermic reaction. Show how the curve changes when the reaction occurs in the presence of a catalyst. Explain how a catalyst increases the rate of a chemical reaction in terms of the meaning of your plot.

7. Samples of gaseous nitrogen and hydrogen are injected into an otherwise empty flask and allowed to react, forming gaseous ammonia, until equilibrium is reached.

   a. From the time the reactant gases are initially injected into the flask until equilibrium is established, in what direction do the concentrations of nitrogen, hydrogen, and ammonia change: increase, decrease, or stay the same? Explain.

   b. From the time equilibrium is established until one hour after equilibrium is established, in what direction do the concentrations of nitrogen, hydrogen, and ammonia change: increase, decrease, or stay the same? Explain.

   c. Exactly what is equal when equilibrium is established? List some quantities that are not necessarily equal when the system reaches equilibrium.

   d. The state of equilibrium is often called a dynamic equilibrium. What is the meaning of the term *dynamic*? How does this term apply to equilibria? Explain at both the particulate and the macroscopic level.

8. a. Write the equilibrium constant expression for the general reaction

$$a\,A + b\,B \rightleftharpoons c\,C + d\,D$$

b. Define all the terms in your expression.

c. Write the equilibrium constant expression for the reaction

$$4\ NH_3(g) + 5\ O_2(g) \rightleftharpoons 4\ NO(g) + 6\ H_2O(g)$$

d. Equal quantities of hydrogen and iodine gases at 700 K are placed in a reaction vessel and allowed to come to equilibrium according to the equation

$$H_2(g) + I_2(g) \rightleftharpoons 2\ HI(g)$$

The following equilibrium concentrations are measured:

$$[H_2] = [I_2] = 0.021\ \text{mol/L}$$
$$[HI] = 0.158\ \text{mol/L}$$

What is the equilibrium constant for this reaction at 700 K?

9. Student A claims that reactions with relatively fast rates have large K values because they move to equilibrium faster. Student B says that reactions with relatively slow rates have large K values because a greater concentration of products will eventually build up if the reaction is given sufficient time. Who is correct? Explain.

10. a. If the reaction

glucose 6-phospate $\rightleftharpoons$ fructose 6-phosphate

is initially at equilibrium and some fructose 6-phosphate is removed, what will happen to the equilibrium? How will the concentrations of glucose 6-phosphate compare before and after the fructose 6-phosphate is removed?

b. The reaction

sucrose + Pi $\rightleftharpoons$ glucose 1-phosphate + fructose

is allowed to come to equilibrium. Bacterial sucrose phosphorylase is then added to catalyze the reaction. Explain the effect of the catalyst on the relative concentrations of both reactants and products before and after its addition.

c. The reaction

$$\text{glutamate} + \text{oxaloacetate} \rightleftharpoons \alpha\text{-ketoglutarate} + \text{aspartate}$$

is used by cells to control the quantity of amine groups for amino acids. Explain how the equilibrium shifts (i) when the concentration of glutamate increases and (ii) when the concentration of α-ketoglutarate increases.

d. Reactions involving adenosine 5-triphosphate, or ATP, are used to transfer energy within cells. One important reaction is

$$ATP^{4-} + H_2O \rightleftharpoons ADP^{3-} + HPO_4^{2-} + H^+ + 30\text{ kJ}$$

Chemists often study reactions at the thermodynamic standard temperature of 25°C. How would the equilibrium concentrations of reactants and products differ for this reaction at body temperature, 37°C?

11. In this exercise, your Workshop group will model the particulate-level behavior of a reversible reaction with the goal of understanding how concentration changes with time for an equilibrium reaction. Work in groups of three, with one person representing the reacting molecules, another representing the product molecules, and the third recording the data. Begin by tearing a sheet of paper into 100 pieces. Each piece of paper represents one molecule.

The reaction you will model is A $\rightleftharpoons$ B, and when a piece of paper (molecule) is on the reactant side of the desk, it represents an A molecule, and it is a molecule of B when it is on the product side of the desk. Each step in this exercise will represent one second of time. The group member responsible for recording the data will write down the number of A molecules and the number of B molecules after each step (second) in the exercise in the following table.

Time (s)	Number of A molecules	Number of B molecules
0		
1		
2		
3		
4		
5		
6		
7		
8		
9		
10		
11		
12		
13		
14		
15		

In each step (representing one second) allow 10% of the A molecules to change to B molecules and then react 10% of the B molecules back to A. Record the number of each after each step. Continue the exercise until you have completed 15 steps.

Assuming that the number of molecules is proportional to the concentration, use your data to draw a plot of concentration vs. time on the following graph. Use one color for the plot of molecule A and another for B.

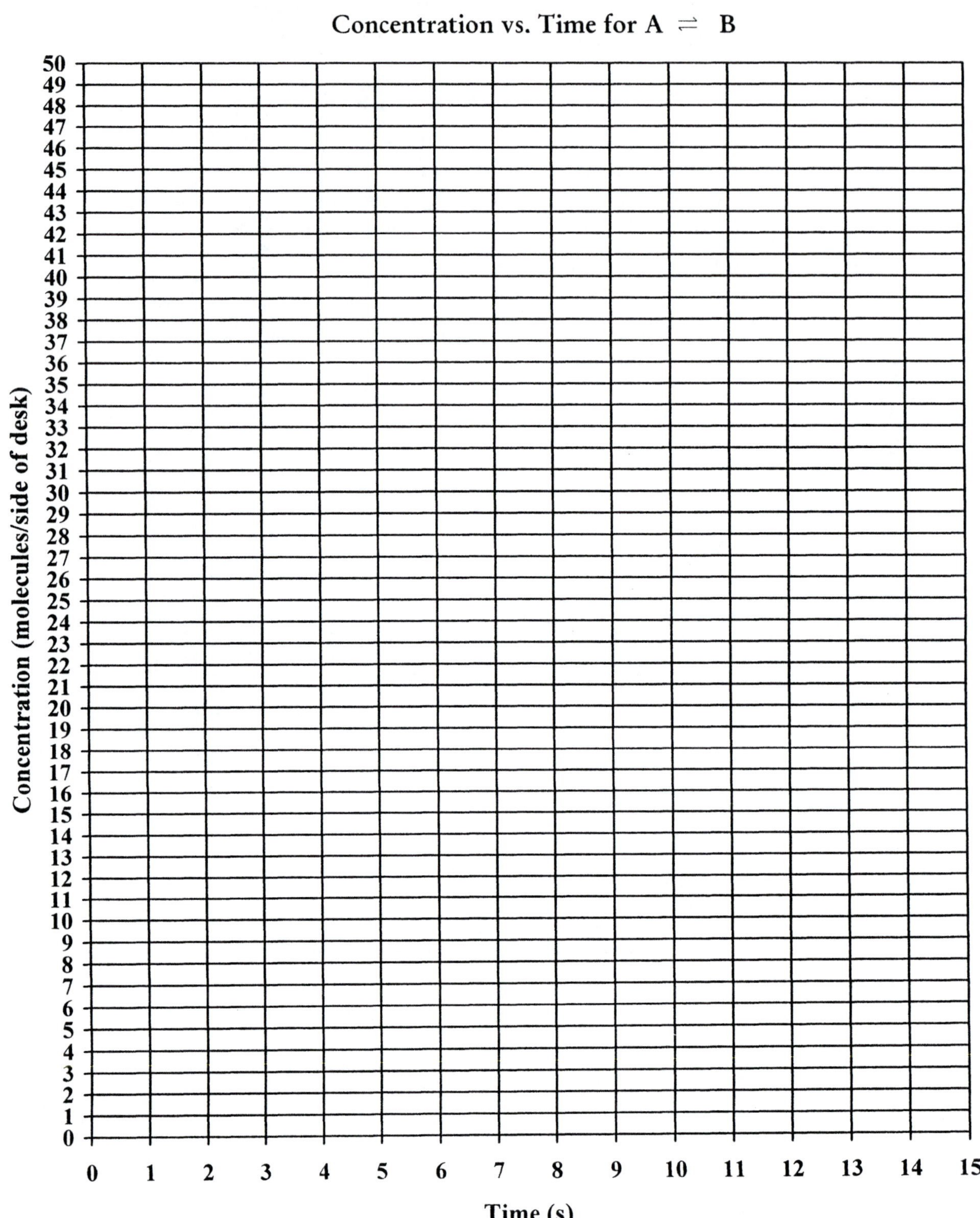

a. When does the net reaction proceed at the fastest rate? When is the rate the slowest? How does this relate to your modeling exercise?

b. How does the reversibility of the reaction affect the concentrations of reactant and product? How would your plot differ if the reaction was not reversible?

c. Did your reaction got to equilibrium? How do you know when equilibrium is achieved?

d. What would a plot of net reaction rate versus time look like for your reaction. Draw a sketch of such a plot and explain its important features.

12. Reconsider your original answers in the *Topics to Review* section. Has your understanding of these topics changed as a result of this Workshop? Summarize your modified answers in the space below.

Acids and Bases

Topics to Review

1. What are the Arrhenius definitions of *acid* and *base*?

   Textbook pages _____ *to* _____.

2. What are the Brønsted–Lowry definitions of *acid* and *base*?

   Textbook pages _____ *to* _____.

3. What distinguishes a strong acid from a weak acid? Give an example of each to support your explanation.

   Textbook pages _____ *to* _____.

4. What is an acid–base neutralization reaction? Give an example of such a reaction.

   Textbook pages _____ *to* _____.

5. What is the self-ionization (or autoionization) of water?

   Textbook pages _____ *to* _____.

6. What is the concentration of hydrogen ion (hydronium ion) in pure water at 25°C? What is the hydroxide ion concentration?

Textbook pages _____ *to* _____.

7. Define pH and pOH.

Textbook pages _____ *to* _____.

8. What is an *indicator*?

Textbook pages _____ *to* _____.

Questions

1. Define the terms *acid* and *base*. Compare the macroscopic definitions with the particulate-level definitions. Give two examples of each that can be found in your local discount and/or grocery store.

2. What is the meaning of *strong* and *weak* acids and bases? Are these terms absolutes or two ends of a spectrum? Is a strong acid more corrosive than a weak acid?

3. Consider the following acids and conjugate bases, listed in order of decreasing acid strength:

Acid Formula		Base Formula		
HI	$\rightleftharpoons$	H^+	+	I^-
H_2SO_4	$\rightleftharpoons$	H^+	+	HSO_4^-
$H_2C_2O_4$	$\rightleftharpoons$	H^+	+	$HC_2O_4^-$
HF	$\rightleftharpoons$	H^+	+	F^-
$HC_2H_3O_2$	$\rightleftharpoons$	H^+	+	$C_2H_3O_2^-$
H_2S	$\rightleftharpoons$	H^+	+	HS^-
H_3BO_3	$\rightleftharpoons$	H^+	+	$H_2BO_3^-$
HPO_4^{2-}	$\rightleftharpoons$	H^+	+	PO_4^{3-}

a. Give the Brønsted–Lowry definitions of the following terms: *acid, base, acid–base reaction.*

b. Write the formula of the conjugate base of NH_4^+.

c. Write the formula of the conjugate acid of CO_3^{2-}.

d. List the following acids in order of decreasing strength: H_2SO_4, HF, H_2S.

e. List the following bases in order of decreasing strength: HSO_4^-, F^-, HS^-.

f. Explain the relationship between your answers to parts (d) and (e).

g. Write the net ionic equation for the reaction between each of the following and predict which side will be favored at equilibrium:

HI and $HC_2O_4^-$

H_3BO_3 and $C_2H_3O_2^{2-}$

4. What is a *neutralization reaction*? Give two examples.

5. Without using a calculator, fill in the blanks in the following table:

pH	pOH	[H+]	[OH−]	Acidic, Basic Neutral
3	11	10^{-3}	10^{-11}	acidic
6	8	10^{-6}	10^{-8}	acidic
10	4	10^{-10}	10^{-4}	basic
9	5	10^{-9}	10^{-5}	basic

Use a calculator to fill in the blanks in the following table:

pH	pOH	[H+]	[OH−]	Acidic, Basic Neutral
8.12				
	3.44			
		3.4×10^{-9}		
			1.2×10^{-7}	

6. What are acid–base indicators? Why are they used? Give two examples of indicators.

7. a. List the formulas of the common strong acids.

b. List the formulas of the common strong bases.

c. Will the solution formed from dissolving each of the following ionic salts in water be acidic, basic, or neutral?

i. A salt of a strong acid and a weak base

ii. A salt of a strong base and a weak acid

iii. A salt of a strong acid and a strong base

neurral

d. Classify solutions formed from dissolving the following ionic salts in water as acidic, basic, or neutral. Explain how each ion of each salt affects the solution pH.

i. Na_3PO_4

ii. $Al(NO_3)_3$

iii. KCl

8. a. What is the *self-ionization of water*? Write the equation that describes the process.

b. What is meant by the symbol K_w? What is its value? What is the relationship between K_w and temperature?

9. a. Sodium carbonate, Na_2CO_3, and ammonia, NH_3, are among the most commonly used bases in both academic and industrial settings. Neither compound contains the hydroxide ion. How then can they be bases? Discuss this questions with your group and explain.

b. Write the equation for the reaction of HCl with Na_2CO_3, with NH_3, and with NaOH.

10. Brainstorm to come up with a list of concepts related to acids, bases, and acid–base reactions. Summarize each concept with a single term or brief phrase. Define each of these phrases. Enclose each term or phrase with a rectangle. Link rectangles with lines, and write a brief description of the relationship between the concepts on each line.

11. Reconsider your original answers in the *Topics to Review* section. Has your understanding of these topics changed as a result of this Workshop? Summarize your modified answers in the space below.

Buffers and Titrations

Topics to Review

1. What is a buffer solution?

   Textbook pages _____ *to* _____.

2. Which of the following can be combined to form a buffer solution: strong acid, weak acid, strong base, weak base? Explain.

   Textbook pages _____ *to* _____.

3. How does a buffer system resist changes in pH when acid or base is added?

   Textbook pages _____ *to* _____.

4. If weak acid and conjugate base concentrations are equal in a buffer solution, what is the relationship between the solution pH and the pK_a of the acid?

   Textbook pages _____ *to* _____.

5. What is meant by the term *buffer capacity*?

   Textbook pages _____ *to* _____.

6. Describe with words and a sketch how and why a titration is performed.

Textbook pages _____ *to* _____.

7. What is an *indicator*?

Textbook pages _____ *to* _____.

8. Explain how to determine the equivalent mass of an acid or a base.

Textbook pages _____ *to* _____.

9. What is the advantage of using equivalents when performing acid–base neutralization reactions?

   Textbook pages _____ *to* _____.

10. Write the mathematical definition of normality. What is the relationship between normality and molarity?

    Textbook pages _____ *to* _____.

Questions

1. A buffer is prepared from equal molar quantities of acetic acid, $HC_2H_3O_2$, and sodium acetate, $NaC_2H_3O_2$. Explain how the buffer keeps the pH of the solution at about the same value (a) after the addition of a few drops of a dilute HCl solution and (b) after the addition of a few drops of a dilute NaOH solution. Use reaction equations to describe the chemical changes that occur.

2. a. How would your answers to Question 1 differ if the buffer was prepared from equal molar quantities of sodium bicarbonate, $NaHCO_3$, and sodium carbonate, Na_2CO_3?

   b. How would your answers compare if the buffer in Question 1 (i) was in a solution with a total volume of 1.0 L versus (ii) a total solution volume of 2.0 L?

3. The following acids and soluble salts of their conjugate bases are available in a laboratory:

Acid	pK_a
Acetic	4.74
Formic	3.74
Hypochlorous	7.52
Iodic	0.77

Which combination would be best to use to prepare a buffer at (a) pH = 5, (b) pH = 7? Explain how you made your choices.

4. A buffer solution is made from 1.00 mol of a weak acid, HA, and 1.00 mol of the sodium salt of its conjugate base, NaA.

a. A total of 1.50 mol of a strong acid, HCl, is added to the buffer solution in 0.25-mol increments. If the buffer system is initially at pH = 7.0 and the acid is added with no change in volume, how will the pH of the solution change? Answer this question by sketching a qualitative plot of pH (*y*-axis) versus moles of acid added (*x*-axis).

b. Sketch a plot similar to that in part (a) showing how the pH of the buffer solution would change with the addition of 1.50 mol of a strong base, NaOH, in 0.25-mol increments.

c. What is meant by *buffer capacity*? How are your results from parts (a) and (b) related to the answer to this question?

5. The buffer system primarily responsible for maintaining constant pH in blood serum and extracellular fluids is usually referred to as the bicarbonate buffer system:

$$CO_2(g) + H_2O(\ell) \rightleftharpoons H_2CO_3(aq) \rightleftharpoons H^+(aq) + HCO_3^-(aq)$$

If we consider $CO_2(g) + H_2O(\ell)$ as essentially the same as $H_2CO_3(aq)$, the Henderson–Hasselbalch equation is:

$$pH = pK_a + \log \frac{[HCO_3^-]}{[CO_2]}$$

where $K_a = 4.3 \times 10^{-7}$ for $H_2CO_3(aq)$.

a. What is the ratio $[HCO_3^-]/[CO_2]$ at a normal blood pH of 7.4?

b. The concentration of HCO_3^- in blood is normally 0.025 M. What is the concentration of CO_2?

c. How do you suppose breathing is related to this buffer system? What initially happens to your rate of breathing when the CO_2 concentration in the blood is too high?

6. The major buffer system within cells is usually referred to as the phosphate buffer system:

$$H_2PO_4^-(aq) \rightleftharpoons H^+(aq) + HPO_4^{2-}(aq)$$

where $K_a = 6.2 \times 10^{-8}$ for $H_2PO_4^-(aq)$.

a. Write the Henderson-Hasselbalch equation for this buffer system.

b. What ratio of $[HPO_4^{2-}]$ to $[H_2PO_4^-]$ is necessary to create a solution buffered at pH = 7.4?

c. What quantities of NaH_2PO_4 and Na_2HPO_4 are required to make a pH = 7.4 buffer solution that has the capacity to resist the addition of 0.10 mol of strong acid or base?

7. Write a description of what a titration is and why it is performed so that a person without a background in chemistry can understand the technique. Use sketches as necessary to supplement your written description.

8. a. The term *equivalent* means "of equal value" in everyday language. How is this term defined in chemistry? How do the everyday and chemistry definitions compare? Why do chemists use equivalents?

b. Determine the equivalent mass of each of the following: HCl, H_2SO_4, KOH, $Ba(OH)_2$.

c. Write a brief procedure explaining how to determine the equivalent mass of any acid or base.

9. a. Define the term *normality*, as used in chemistry.

b. Explain why the formula $V_{acid}N_{acid} = V_{base}N_{base}$ can be used to determine an unknown normality in an acid–base titration.

c. How is the formula $V_{acid}N_{acid} = V_{base}N_{base}$ affected when the acid used in the titration is H_2SO_4 instead of HCl? How is the formula affected when the base is $Mg(OH)_2$ instead of NaOH? Explain your answers to these questions.

10. Your group is to prepare an HCO_3^- solution that approximates the normal concentration in blood, which is 24 meq/L. Give a *complete* description of how you would make 10.0 L of this solution.

11. Consider the titration of 0.100 M NaOH into 50.0 mL of a 0.100 M HCl solution.

a. The following chart contains entries for various amounts of added NaOH solution. Complete the chart by calculating the values for each column. You may leave blank the boxes with a dash in them. The first two lines are completed as examples.

mL NaOH added	mmol OH^- added	mmol H^+ in solution	mmol OH^- in solution	Total volume (mL)	$[OH^-]$	$[H^+]$	pH
0.0	0.0	5.00	—	50.0	—	1.00×10^{-1}	1.00
10.0	1.00	4.00	—	60.0	—	6.67×10^{-2}	1.17
20.0			—		—		
30.0			—		—		
40.0			—		—		
49.0			—		—		
49.9			—		—		
50.0			—		—		
50.1							
51.0							
60.0							
70.0							

b. Construct a plot of pH (y-axis) versus milliliters of 0.100 M NaOH added (x-axis), using the data from the chart. Identify what is occurring chemically in each section of the curve.

12. Reconsider your original answers in the *Topics to Review* section. Has your understanding of these topics changed as a result of this Workshop? Summarize your modified answers in the space below.

Oxidation–Reduction

Topics to Review

1. Define the terms *oxidation* and *reduction* in terms of both electron transfer and oxidation number change.

   Textbook pages _____ *to* _____.

2. Write a brief summary of the procedure used to assign oxidation numbers.

   Textbook pages _____ *to* _____.

3. Explain the difference between a *voltaic* (galvanic) and an *electrolytic* cell.

   Textbook pages _____ *to* _____.

4. Compare the terms *oxidizing agent* and *reducing agent*.

   Textbook pages _____ *to* _____.

Questions

1. a. Define the words *oxidation* and *reduction* in terms of each of the following.

   i. The gain or loss of oxygen atoms

   ii. The gain or loss of hydrogen atoms

   iii. The gain or loss of electrons

   b. Which definition is the most generalizable? Explain.

   c. Under what conditions can oxidation occur without reduction and vice versa?

2. a. Define each and state the relationships among the following terms.

   i. Strong oxidizing agent

   ii. Weak oxidizing agent

   iii. Strong reducing agent

   iv. Weak reducing agent

   b. When a strong oxidizing agent gains an electron, it becomes a weak reducing agent. Explain.

   c. Your group is given strips of copper and lead and solutions of copper(II) nitrate and lead(II) nitrate. Your task is to determine which is the stronger oxidizing agent, copper(II) ion or lead(II) ion. Explain how you would conduct an experiment to complete your task. Would the results from this experiment also provide evidence about strengths of reducing agents? Explain.

3. Use your knowledge of oxidation–reduction reactions and your everyday life experiences to classify each of the following as an oxidizing agent or a reducing agent. In each case, explain the reasoning you used to come to your conclusion.

Household bleach

Carbon in the reaction $SnO_2 + C \rightarrow Sn + CO_2$

Oxygen

Hydrogen

Vitamin C (ascorbic acid)

Antiseptics (such as a 3% solution of hydrogen peroxide)

Disinfectants (such as swimming pool chlorine)

4. a. What is an antioxidant?

   b. What would a chemist call an antioxidant?

   c. What are common examples of antioxidants?

   d. How does your answer to this question relate to Questions 1 and 2?

5. Consider each of the following facts about acid–base reactions, and then by analogy, explain how oxidation–reduction reactions have similar characteristics. Part (a) has been completed as an example.

   a. Acid-base reactions are proton-transfer reactions.

      Oxidation–reduction reactions are electron-transfer reactions.

   b. The terms *strong* and *weak* are used to classify acids and bases on their abilities to donate and accept protons.

   c. Some species (such as water) can both accept and donate protons.

   d. The favored side of an equilibrium in an acid–base reaction can be predicted by comparing acid–base strengths.

6. Consider the following reactions, which are the reverse of each other:

$$C_6H_{12}O_6 + 6\,O_2 \rightarrow 6\,CO_2 + 6\,H_2O + \text{energy}$$
$$6\,CO_2 + 6\,H_2O + \text{energy} \rightarrow C_6H_{12}O_6 + 6\,O_2$$

a. What is the role of these reactions in living organisms? Discuss this question and explain in as much detail as possible.

b. Are these oxidation–reduction reactions? Explain.

c. How does the movement of electrons to higher or lower energies correlate with the energy changes in these reactions?

d. For what is the energy from the first reaction used? What is the source of the energy in the second reaction? Explain.

7. a. Develop a procedure that could be followed by a beginning chemistry student for assigning oxidation numbers to elements in a formula.

b. Test your procedure by using your list of rules to determine the oxidation numbers of the underlined elements in the following species. Modify your procedure as necessary if you find that it does not work properly for these formulas.

$\underline{Cl}_2$ $\underline{Zn}^{2+}$ $Na_2\underline{O}$

$H_2\underline{O}_2$ $K\underline{Br}$ $\underline{S}O_4^{2-}$

$\underline{N}O_2$ $\underline{N}O_3^-$ $\underline{N}H_4^+$

$K\underline{Br}O_3$ $\underline{Ba}(OH)_2$

8. a. How do you recognize a chemical change as an oxidation–reduction reaction?

b. Which of the following are oxidation–reduction reactions? Explain how you made your determination in each case.

i. $NaCl(aq) + AgNO_3(aq) \rightarrow NaNO_3(aq) + AgCl(s)$

ii. $CH_4(g) + 2\ O_2(g) \rightarrow CO_2(g) + 2\ H_2O(\ell)$

iii. $2\ HCl(aq) + Zn(s) \rightarrow H_2(g) + ZnCl_2(aq)$

iv. $2\ HBr(aq) + Na_2CO_3(aq) \rightarrow CO_2(g) + H_2O(\ell) + 2\ NaBr(aq)$

9. The lead–acid storage battery found in automobiles is one of the most common applications of oxidation–reduction reactions. When the battery is providing energy to start the engine, the reaction that occurs at the anode is

$$Pb(s) + HSO_4^-(aq) \rightarrow PbSO_4(s) + H^+(aq) + 2\,e^-$$

and the reaction that occurs at the cathode is

$$PbO_2(s) + 3\,H^+(aq) + HSO_4^-(aq) \rightarrow 2\,PbSO_4(s) + 2\,H_2O(\ell)$$

The battery is made from lead, which serves as the anode, and lead(IV) oxide, which serves as the cathode. Sulfuric acid solution, which dissociates to $H^+(aq)$ and $HSO_4^-(aq)$, is the electrolyte.

a. Write the overall reaction.

b. What species is oxidized, and what species is reduced?

c. What is the oxidizing agent? What is the reducing agent?

d. What reaction occurs after the car has been started and the alternator recharges the battery? Explain.

e. How are the chemical change and resulting energy change in a lead–acid storage battery similar to the biological processes in Question 6? Explain.

10. Draw a sketch of a voltaic (galvanic) cell composed of one compartment with a strip of zinc in a solution of zinc sulfate and another compartment with a strip of copper in a copper(II) sulfate solution. Link the solutions in the two compartments with a salt bridge, and link the metal strips with a voltmeter. Label the anode and cathode and indicate the direction of electron flow. Explain how electrons move through the system. Write each half-reaction and the net ionic equation for the overall reaction.

11. Draw a sketch illustrating the electrolysis of liquid sodium chloride. Use inert electrodes, one connected to each side of a battery, immersed in the liquid. Label the anode and cathode and indicate the direction of electron flow. Explain how electrons move through the system. Write each half-reaction and the net ionic equation for the overall reaction.

12. Reconsider your original answers in the *Topics to Review* section. Has your understanding of these topics changed as a result of this Workshop? Summarize your modified answers in the space below.

Structure and Nomenclature

Molecular models will be necessary for this unit.

Topics to Review

1. What is the difference between a structural formula and molecular formula?

   Textbook pages _____ *to* _____.

2. What are isomers?

   Textbook pages _____ *to* _____.

3. What are the four types of tetrahedral carbon?

   Textbook pages _____ to _____.

4. What are the names of the first ten straight chain alkanes?

   Textbook pages _____ to _____.

5. What are the names and structures of all the alkyl groups containing up to four carbons?

   Textbook pages _____ to _____.

6. What are functional groups? What are the names and structures of the most common functional groups?

Textbook pages _____ *to* _____.

7. What are the IUPAC rules of nomenclature for hydrocarbons?

Textbook pages _____ *to* _____.

8. What are geometric isomers?

Textbook pages _____ *to* _____.

Questions

1. Classify the following pairs of compounds as being structural isomers, identical, or different compounds.

a. $CH_3-\underset{CH_3}{\underset{|}{C}H}-CH_2-CH_2-\underset{CH_3}{\underset{|}{C}H}-CH_3$

$(CH_3)_2CH(CH_2)_2CH(CH_3)_2$

b. CH_3 CH_3 CH_3 CH_3

c. CH_3 CH_3 CH_3 CH_2CH_3

d. $\overset{CH_3}{\overset{|}{C}H_2}-CH_2-\overset{CH_3}{\overset{|}{C}H}-CH_2-CH_3$

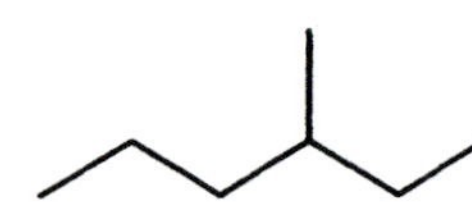

e. CH_3 CH_3

f. 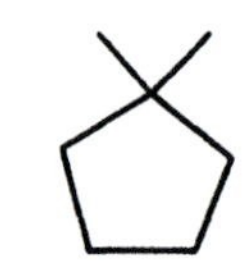

g. 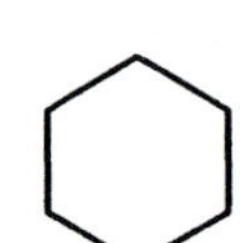 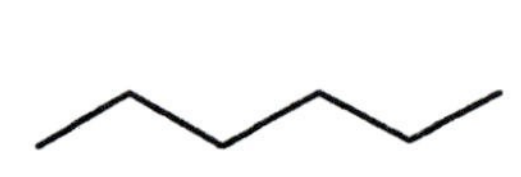

h. 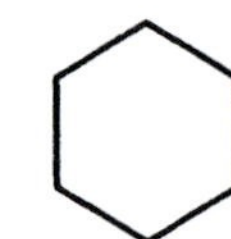 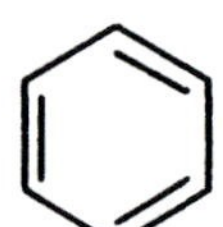

2. Consider the following line structure of a compound:

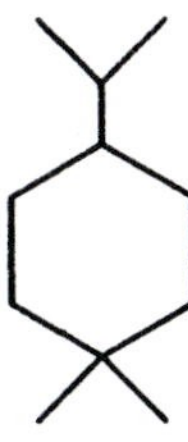

a. What is the expanded structure of this compound?

b. What is the molecular formula of this compound?

c. How many types of carbon does this compound have?

d. Identify each type of carbon and label it on the structure provided.

3. Draw the structures (including geometric isomers) of all the alkenes with the molecular formula $C_2H_2Cl_2$. Give the IUPAC name for each compound you draw.

4. Taxol, isolated from the bark of the Pacific yew tree, is an anticancer drug used in the treatment of ovarian cancer. Circle and name the functional groups present in taxol.

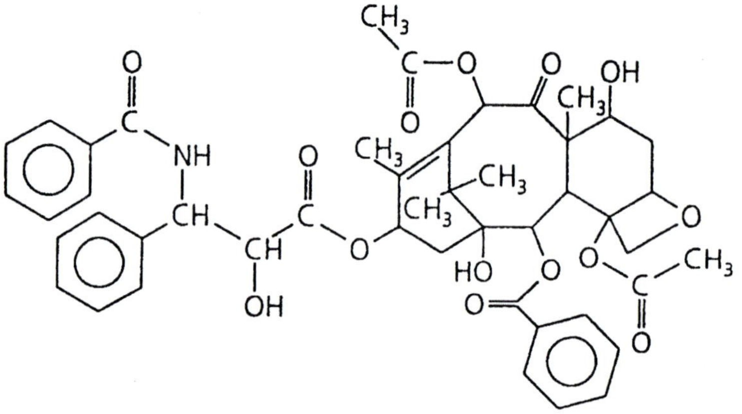

Taxol

5. Are the functional groups in the following two molecules the same? Why or why not?

a. $CH_3CH_2CH_2-\overset{O}{\overset{\|}{C}}-OCH_2CH_3$

b. $CH_3CH_2-\overset{O}{\overset{\|}{C}}-CH_2OCH_2CH_3$

6. Draw structures of molecules with the formula C_4H_8O that contain:

   a. an alcohol

   b. an ether

   c. a ketone

   d. an aldehyde

   e. Can you use this molecular formula to make molecules that contain other functional groups not mentioned in (a) through (d)? If so, draw the structures.

7. Which of the following names are correct and which ones are incorrect? Explain exactly what is wrong with those that are incorrect and rename them correctly using IUPAC rules.

   a. 3-isobutylheptane

   b. 1,1-dichloro-2,3-diphenyl-4,4-dimethylbutane

   c. 2-benzylpentane

   d. 1-methyl-6-isopropylcyclohexane

   e. 2,6-dichloro-4-ethylheptane

   f. 2,6-diiodo-4-ethylheptane

8. Draw and name all possible structural isomers of the dimethylcyclohexanes. Identify the ones that are capable of *cis-trans* isomerism.

9. Reconsider your original answers in the *Topics to Review* section. How has your understanding of these topics changed as a result of this Workshop? Summarize your modified answers in the space below.

Reaction of Hydrocarbons

Topics to Review

1. What are the names of the two most common reactions that alkanes undergo? Write an equation to illustrate each of the reactions.

   Textbook pages _____ *to* _____.

2. Write the general equation for the addition of a symmetrical reagent to an alkene. Name the reactions that belong to this category. In each case, state the reagents and the reaction conditions used.

   Textbook pages _____ *to* _____.

3. Write the equation for the addition of an unsymmetrical reagent to an alkene. Which addition reactions belong in this category? In each case state the reagents and the reaction conditions used.

   Textbook pages _____ *to* _____.

4. What is Markovnikov's rule?

   Textbook pages _____ *to* _____.

5. What type of reaction do most aromatic compounds undergo?

   Textbook pages _____ *to* _____.

Questions

1. The most important chemical property of alkanes is that they burn easily.

   a. Discuss how this property of alkanes is used by most of the world's population.

   b. Write and balance the complete combustion reaction for butane.

   c. Write and balance two incomplete combustion reactions for butane.

   d. Are any of the products of combustion lethal when inhaled? Which ones? Why?

   e. What is the greenhouse effect? Should we be concerned about it? Why?

   f. Which product(s) of alkane combustion are responsible for the greenhouse effect? How does it (they) bring about this effect? Explain clearly to the group.

   g. Discuss ways by which we can decrease the threat of the greenhouse effect.

2. Consider the following reactions. Each member of the Workshop group should name one reaction and explain to the rest of the group how the structure of the starting material and reaction conditions affect the outcome of the reaction.

a. cyclohexane + Br_2 ⟶ No reaction

b. cyclohexene + Br_2 ⟶ 1,2-dibromocyclohexane (Br, Br)

c. benzene + Br_2 ⟶ No reaction

d. benzene + Br_2 $\xrightarrow{FeCl_3}$ bromobenzene (Br) + HBr

e. cyclohexene + H_2 $\xrightarrow[\text{heat/pressure}]{\text{Pt}}$ cyclohexane

f. benzene + H_2 $\xrightarrow[\text{200 °C/pressure}]{\text{Pt}}$ cyclohexane

As a group summarize what you have learned about the relationship between structure and reactivity of hydrocarbons.

3. Complete the following reactions. If more than one product is formed, indicate which is the major product.

a. ? + $2\ H_2$ ⟶ $CH_3CH_2CH_2CH_2CH_3$

b. $CH_3CH_2CH_2C{\equiv}CH$ + Br_2 ⟶ ?

c. $CH_3{-}C(CH_3){=}CH_2$ $\xrightarrow{H_3O^+}$? (major) + ? (minor)

d. ? + Br_2 ⟶ $CH_3{-}C(Br)(CH_3){-}CH_2{-}Br$

e. $CH_3{-}C(CH_3){=}CH_2$ + HCl ⟶ ? (major) + ? (minor)

f. ? $\xrightarrow{H_3O^+}$ 1-methylcyclohexanol (ring carbon bearing CH_3 and OH) (major) + 2-methylcyclohexanol (CH_3 and OH on adjacent ring carbons) (minor)

g. ? $\xrightarrow{H_3O^+}$ $CH_3{-}CH(OH){-}CH(CH_3){-}CH_3$ + $CH_3{-}CH_2{-}C(OH)(CH_3){-}CH_3$

h. 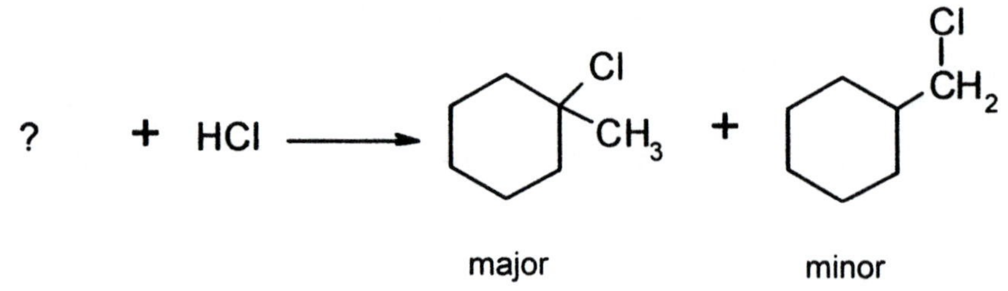 ? + HCl ⟶ (1-chloro-1-methylcyclohexane, CH_3, Cl) + (cyclohexyl-CH_2-Cl)

major minor

4. Complete the following reactions.

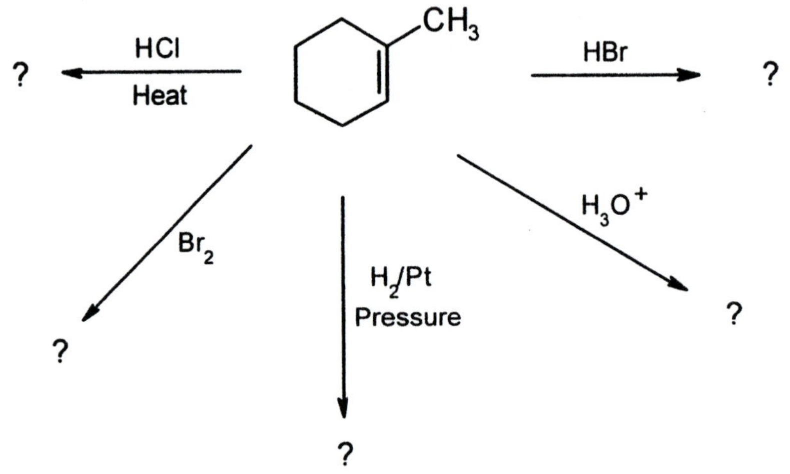

? ⟵ HCl / Heat — 1-methylcyclohexene (CH_3) — HBr ⟶ ?

Br_2 ⟶ ?

H_2/Pt, Pressure ⟶ ?

H_3O^+ ⟶ ?

5. Reconsider your original answers in the *Topics to Review* section. How has your understanding of these topics changed as a result of this Workshop? Summarize your modified answers in the space below.

Alcohols, Phenols, and Ethers

Topics to Review

1. How are alcohols classified?

   Textbook pages _____ *to* _____.

2. What is the difference between alcohols and phenols?

   Textbook pages _____ *to* _____.

3. How does the structure of alcohols affect their physical properties?

   Textbook pages _____ *to* _____.

4. How does the structure of ethers affect their physical properties?

   Textbook pages _____ *to* _____.

5. How do alcohols and phenols react with an oxidizing reagent?

Textbook pages _____ *to* _____.

6. What is the general reaction for dehydration of an alcohol?

Textbook pages _____ *to* _____.

Questions

1. Classify the following as a primary, secondary, or tertiary alcohol or phenol.

a. C_6H_5–CH_2OH (benzene ring–CH_2OH)

b. CH_3–C(CH_3)(CH_3)–CH_2OH

c. CH_3–CH(OH)–CH(NH_2)–COOH

d. CH_3–C(H)(OH)–CH_3

e. cyclohexane ring–OH

f. HO–(benzene ring)–CH_2–CH(NH_2)–COOH

g. cyclohexane ring with CH_3 and OH on the same carbon

h. benzene ring–C(=O)–OCH_3, with OH on the adjacent ring carbon

i. HO–CH_2–CH(NH_2)–COOH

2. Can you have a quaternary alcohol? Explain your answer clearly.

3. Show all the hydrogen bonds that will form in:

   a. pure ethanol

   b. an aqueous solution of ethanol

   c. an aqueous solution of dimethyl ether

   d. Explain why dimethyl ether is completely soluble in water but diethyl ether is only slightly soluble in water, even though they both form hydrogen bonds with water.

4. Explain why methanol with a molecular mass of 32 g/mol is a liquid at room temperature but propane with a molecular mass of 44 g/mol is a gas at room temperature.

5. Specify the compound in each of the following pairs that is more soluble in water. Explain your choice to the group in words and with structures where applicable.

a. $CH_3(CH_2)_5CH_2OH$ and $CH_3CH_2CH_2CH_2CH_2OH$

b. $CH_3OCH_2CH_3$ and $CH_3CH_2CH_2OH$

c. OH and OH

d. CH_3OH and CH_3Cl

e. $CH_3CH_2OCH_2CH_3$ and CH_3OCH_3

6. Arrange the following in (a) order of increasing boiling points and (b) order of increasing viscosity (resistance to flow). Use words and structures to justify your answers.

i. $HOCH_2CH_2OH$

ii. $CH_3CH_2CH_2OH$

iii. $\underset{|}{CH_2}-\underset{|}{CH}-\underset{|}{CH_2}$ with OH on each carbon: $CH_2(OH)-CH(OH)-CH_2(OH)$

7. An important type of biochemical reaction is the oxidation of alcohols. In cells, these reactions are catalyzed by enzymes. In the laboratory, they are carried out by reacting the alcohols with acidified dichromate. Write the structure of the organic products that form when each of the following alcohols reacts with potassium dichromate and sulfuric acid.

a. Cyclohexyl–CH_2OH $\xrightarrow[\text{heat}]{K_2Cr_2O_7/H_2SO_4}$?

b. Cyclohexyl–OH $\xrightarrow[\text{heat}]{K_2Cr_2O_7/H_2SO_4}$?

c. 1-(CH_3)-cyclohexyl–OH $\xrightarrow[\text{heat}]{K_2Cr_2O_7/H_2SO_4}$?

8. Another important type of biochemical reaction is the dehydration of alcohols. Give the reagents and reaction conditions you would use to carry out these reactions with the following alcohols. In each case write the structure(s) of the organic product(s) formed.

a. (1-methylcyclohexanol: cyclohexane ring with CH_3 and OH on the same carbon) ⟶

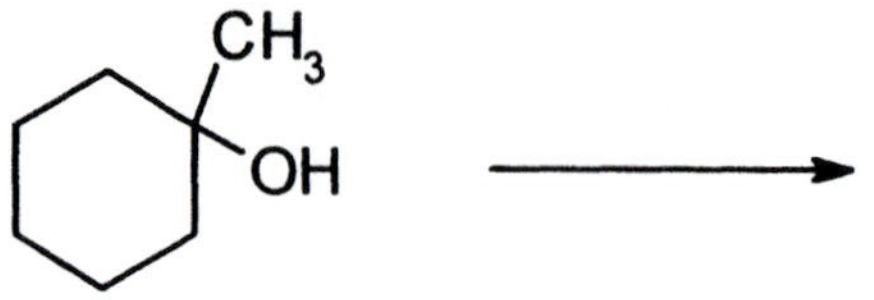

b. (cyclohexane ring with CH_2OH) ⟶

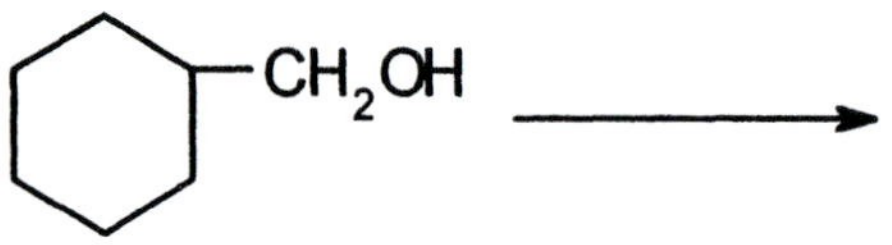

c. (cyclohexane ring with OH) ⟶

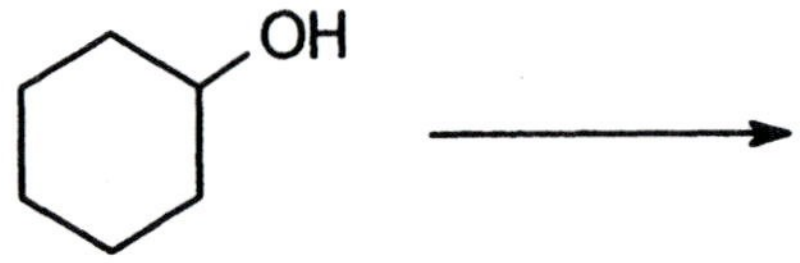

9. Use words and equations to explain the observations in the following reactions:

a. 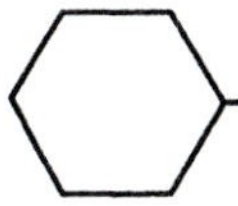 Cyclohexanol (cyclohexyl–OH) + NaOH (aq) ⟶ insoluble

b. 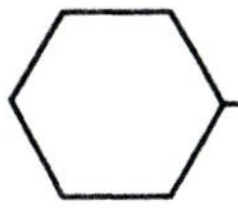 Cyclohexanol (cyclohexyl–OH) + $NaHCO_3$ (aq) ⟶ insoluble

c.  Phenol (phenyl–OH) + NaOH (aq) ⟶ soluble

d.  Phenol (phenyl–OH) + $NaHCO_3$ (aq) ⟶ insoluble

e.  Benzoic acid (phenyl–COOH) + NaOH (aq) ⟶ soluble

f. 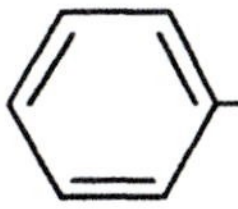 Benzoic acid (phenyl–COOH) + $NaHCO_3$ (aq) ⟶ soluble with formation of gas bubbles

10. Reconsider your original answers in the *Topics to Review* section. How has your understanding of these topics changed as a result of this Workshop? Summarize your modified answers in the space below.

Aldehydes and Ketones

Topics to Review

1. What are the IUPAC rules for naming aldehydes and ketones?

   Textbook pages _____ *to* _____.

2. Write the general reaction for the formation of acetals.

   Textbook pages _____ *to* _____.

3. Write the general reaction for the formation of ketals.

Textbook pages _____ to _____.

4. Write general structures for each of the following. Label each as stable or unstable.

Textbook pages _____ to _____.

Hemiacetal

Acetal

Hemiketal

Ketal

Questions

1. Identify the functional groups in each of the following compounds.

a.  O

b.  Br; OH

c. 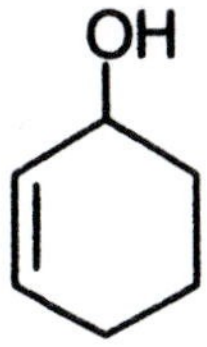 OH

d. 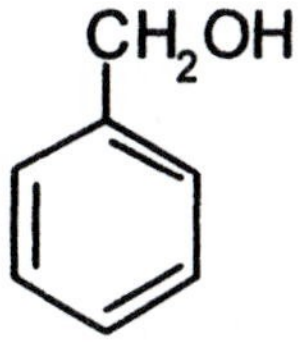 CH_2OH

e. 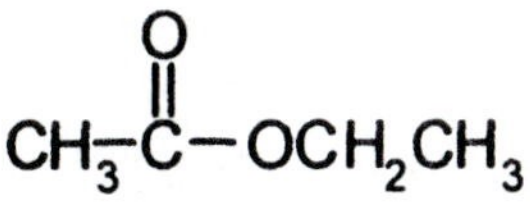 $CH_3-\overset{O}{\overset{\|}{C}}-OCH_2CH_3$

f. 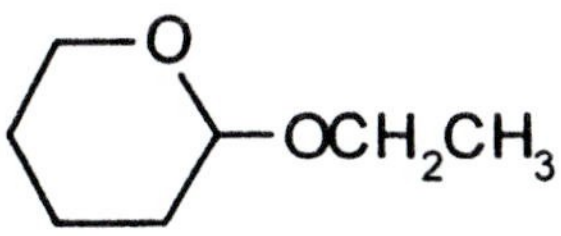 O; OCH_2CH_3

g. 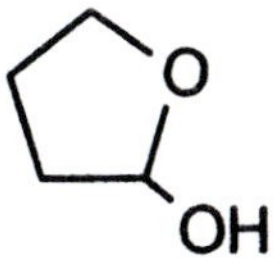 O; OH

h. 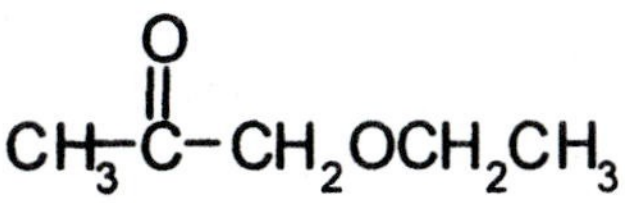 $CH_3-\overset{O}{\overset{\|}{C}}-CH_2OCH_2CH_3$

i. 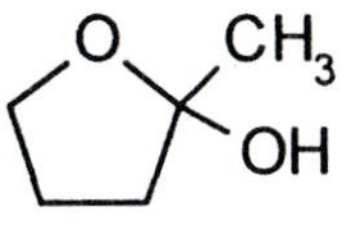 O; CH_3; OH

j. 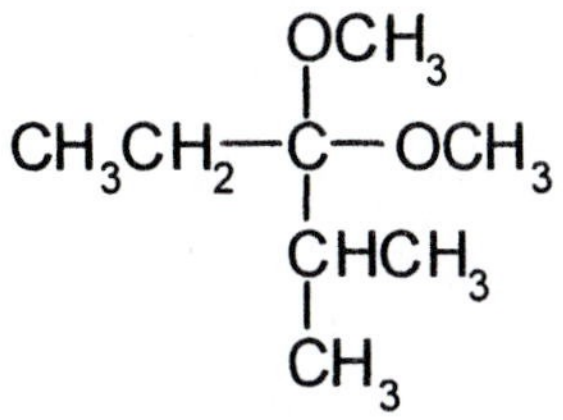 $CH_3CH_2-\underset{\underset{\underset{CH_3}{|}}{CHCH_3}}{\underset{|}{\overset{\overset{OCH_3}{|}}{C}}}-OCH_3$

k. $HOCH_2CH_2OH$

l. 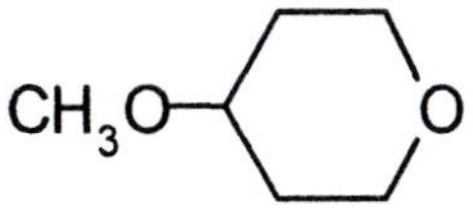 CH_3O; O

m. $CH_3-\underset{\underset{CH_3}{|}}{CH}-\overset{O}{\overset{\|}{C}}-OCH_2CH_3$

n. CH_2OCH_3

o. $\overset{O}{\overset{\|}{C}}-OCH_3$

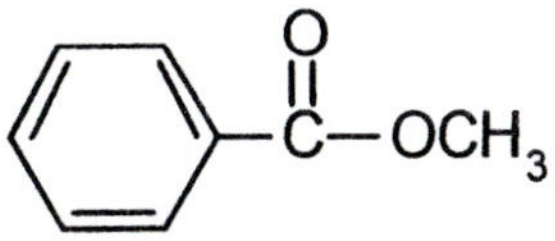

p. 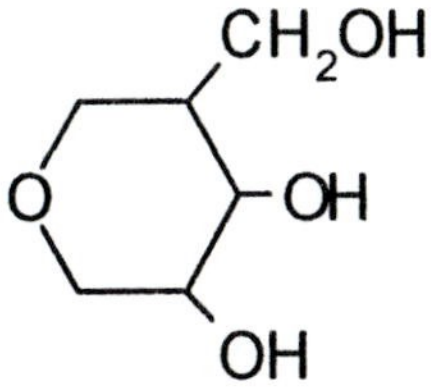 CH_2OH; O; OH; OH

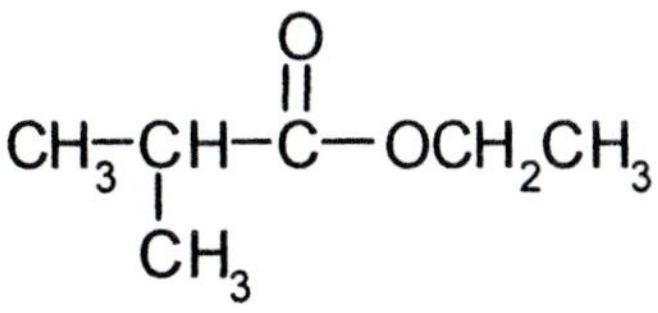

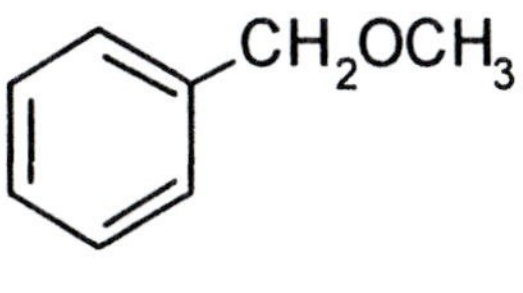

2. Name the following compounds using IUPAC rules.

a. $CH_3CHCH_2CH_2CHO$ with CH_2CH_3 on the second carbon

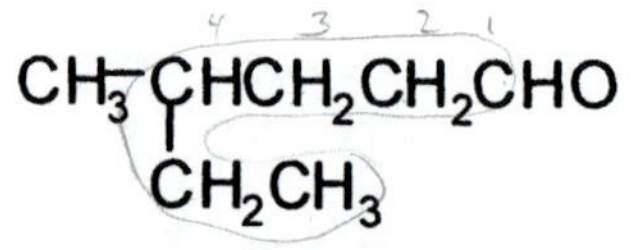

4-methyl hexanal

b. Benzene ring with CH=O, Cl and Br

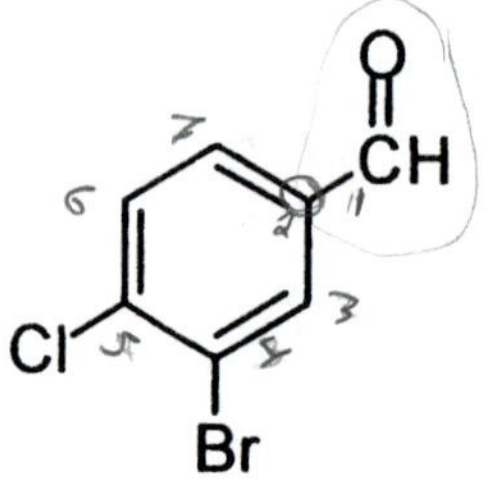

3-bromo 4-chloro cyclohexanal

c. $CH_3-\overset{O}{\overset{\|}{C}}-CHCH(CH_3)_2$ with $CH(CH_3)_2$ on the CH carbon

$CH_3-\overset{O}{\overset{\|}{C}}-CHCHCH_3CH_3$ with $CHCH_3CH_3$ on the CH carbon

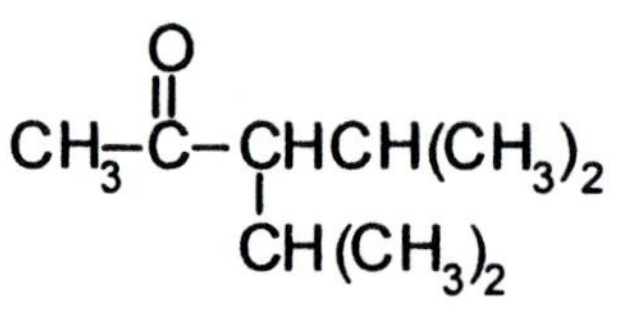

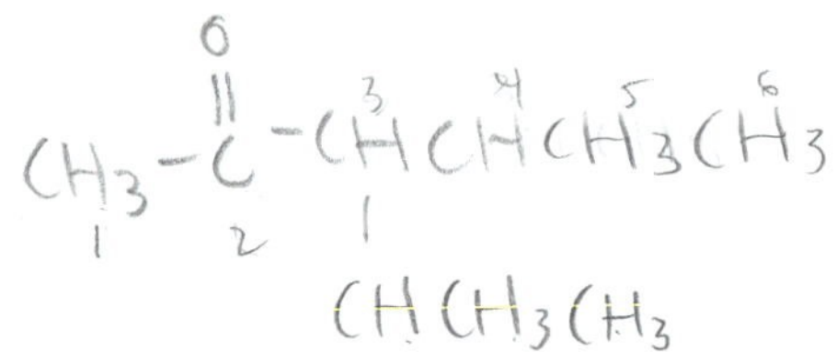

3-propyl 2-hexanal

3. Complete the following reactions:

a. ? + CH_3OH $\rightleftharpoons$ $H_3C-\overset{\displaystyle OH}{\underset{\displaystyle OCH_3}{\overset{|}{\underset{|}{C}}}}-H$

$H_3C\overset{\displaystyle O}{\overset{||}{C}}HOCH_3$

b. $CH_3CH_2-\overset{\displaystyle OH}{\underset{\displaystyle OCH_3}{\overset{|}{\underset{|}{C}}}}-H$ + CH_3CH_2OH $\overset{H_3O^+}{\rightleftharpoons}$? + H_2O

$CH_3CH_2\overset{\displaystyle O}{\overset{||}{C}}OH$

c. ? + ? $\overset{H_3O^+}{\rightleftharpoons}$ (five-membered ring with O; ring carbon bearing CH_3 and OCH_3) + H_2O

d. $CH_3-\overset{\displaystyle O}{\overset{||}{C}}-CH_3$ + 2 CH_3OH $\overset{H_3O^+}{\rightleftharpoons}$? + H_2O

e. ? $\xrightleftharpoons{H_3O^+}$ (cyclic hemiacetal: six-membered ring with O, C bearing H and OH) + H_2O

f. (five-membered ring with O, C bearing CH_3 and OH) $\xrightleftharpoons{H_3O^+}$?

(hint: ring opening)

g. $CH_3-C(OCH_2CH_3)(OCH_3)-CH_2CH_3$ $\xrightleftharpoons{H_3O^+}$? + ? + ?

h. (six-membered ring with O; substituents CH_3O, CH_2OCH_3, CH_3O, CH_3O, OH) $\xrightleftharpoons{H_3O^+}$?

i. (disaccharide: CH_2OH, O, OH, OH, OH, O, $HOCH_2$, O, HO, CH_2OH, OH) $\xrightleftharpoons{H_3O^+}$? + ?

4. The common name for the following aldehyde is chloral.

$$\mathrm{Cl-\underset{\underset{Cl}{|}}{\overset{\overset{Cl}{|}}{C}}-\overset{\overset{O}{\|}}{C}-H}$$

a. What is its IUPAC name for chloral?

b. Chloral reacts with water to form a compound called chloral hydrate, which is used as a sedative. Write the equation for the formation of chloral hydrate.

5. Reconsider your original answers in the *Topics to Review* section. How has your understanding of these topics changed as a result of this Workshop? Summarize your modified answers in the space below.

Carboxylic Acids

Topics to Review

1. What are the IUPAC rules for naming each of the following?

   a. Carboxylic acids

   Textbook pages ___ *to* ___.

   b. Carboxylic esters

   Textbook pages ___ *to* ___.

   c. Carboxylic acid salts

   Textbook pages ___ *to* ___.

2. Write the general reactions for each of the following:

   a. Reaction of carboxylic acids with sodium hydroxide

      Textbook pages ___ *to* ___.

   b. Reaction of carboxylic acids with sodium bicarbonate

      Textbook pages ___ *to* ___.

   c. Preparation of carboxylic acids

      Textbook pages ___ *to* ___.

   d. Preparation of carboxylic esters

      Textbook pages ___ *to* ___.

   e. Hydrolysis of carboxylic esters

      Textbook pages ___ *to* ___.

   f. Saponification of carboxylic esters

      Textbook pages ___ *to* ___.

Questions

1. The reaction of vinegar with baking soda is used by elementary school students to produce "volcanoes." Give the equation for this reaction. Which product is responsible for the volcanic action?

2. When wine is exposed to air for an extended period of time it changes to vinegar. Give the equation for this reaction.

3. Sodium benzoate and calcium propanoate are used as food preservatives. Aluminum acetate is used to treat minor skin irritations. Give the structures of these acid salts.

4. The following compound is generated by the reaction of a carboxylic acid with an alcohol.

$$H_3C-\overset{\displaystyle O}{\overset{\|}{C}}-O-\underset{\displaystyle CH_3}{\underset{|}{\overset{\displaystyle CH_3}{\overset{|}{CH}}}}$$

a. Circle the part of this compound that came from the carboxylic acid. Draw the structure of the acid.

b. Circle the part of this compound that came from the alcohol. Draw the structure of the alcohol.

c. What reaction conditions are needed?

d. Which functional group has been generated?

e. What is the name of this reaction?

f. Is this reaction reversible?

g. Write the complete equation for this reaction.

5. Because of their pleasant odors many esters are used as artificial flavors in food products. Give the structures of the following esters. In each case write the structure of the alcohol and acid used to synthesize the ester.

Ester	Flavor	Ester Structure	Acid Structure	Alcohol Structure
ethyl methanoate	rum			
isopentyl ethanoate	banana			
ethyl butanoate	pineapple			
isobutyl methanoate	raspberry			

6. Write the hydrolysis reaction for the following triglyceride. Give the conditions and products.

$$
\begin{array}{l}
H_2C-O-\overset{\displaystyle O}{\overset{\|}{C}}-C_{17}H_{35} \\
\quad | \\
HC-O-\overset{\displaystyle O}{\overset{\|}{C}}-C_{17}H_{35} \\
\quad | \\
H_2C-O-\overset{\displaystyle O}{\overset{\|}{C}}-C_{17}H_{35}
\end{array}
$$

7. Heroin is less polar than morphine (see structures below). That is why it enters the brain more rapidly and is more potent.

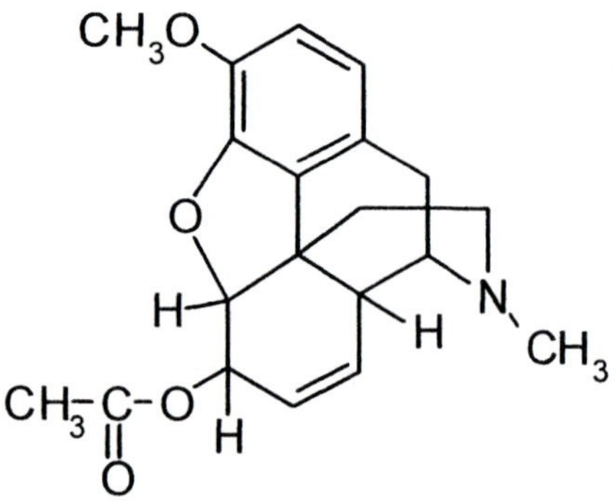

Heroin

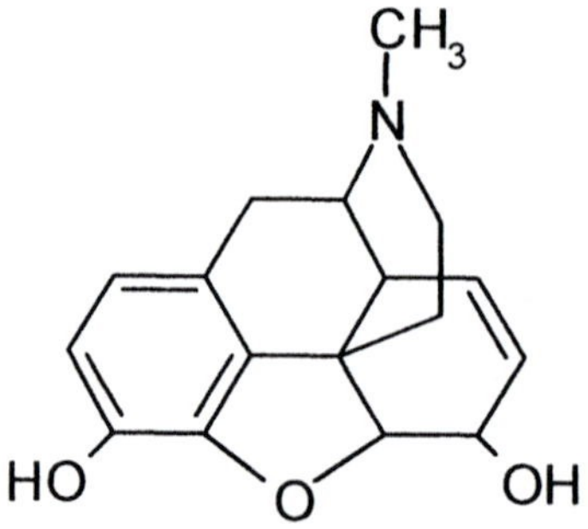

Morphine
an opiod analgesic

a. Why is heroin less polar than morphine?

b. Give the reagents and reaction conditions for conversion of morphine to heroin.

c. Name the reaction that converts morphine to heroin.

8. a. What is saponification?

b. What is the origin of this name?

c. Give the reagents and reaction conditions necessary to carry out saponification.

d. Write the general reaction for a saponification reaction.

9. Write the products of the saponification of the following esters.

a. $C_6H_5-O-\overset{O}{\overset{\|}{C}}-CH_3$

b. $CH_3O-\overset{O}{\overset{\|}{C}}-C_6H_5$

c.

$$
\begin{array}{l}
H_2C-O-\overset{O}{\overset{\|}{C}}-C_{17}H_{35} \\
\;\;| \\
HC-O-\overset{O}{\overset{\|}{C}}-C_{17}H_{35} \\
\;\;| \\
H_2C-O-\overset{O}{\overset{\|}{C}}-C_{17}H_{35}
\end{array}
$$

10. Reconsider your original answers in the *Topics to Review* section. How has your understanding of these topics changed as a result of this Workshop? Summarize your modified answers in the space below.

Amines and Amides

Topics to Review

1. How are amines classified?

   Textbook pages ___ to ___.

2. Compare the way alcohols are classified with the way amines are classified.

   Textbook pages ___ to ___.

3. How does the structure of amines affect their physical properties?

   Textbook pages ___ *to* ___.

4. Write the general equation for the reaction of an amine with an acid.

   Textbook pages ___ *to* ___.

5. How are amine salts formed?

   Textbook pages ___ *to* ___.

6. Write the general equation for the reaction of an amine salt with a base.

Textbook pages ___ to ___.

7. Write the general equation for the formation of an amide.

Textbook pages ___ to ___.

8. Write the general equations for the hydrolysis of amides under:

a. acidic conditions

Textbook pages ___ to ___.

b. basic conditions

Textbook pages ___ to ___.

Questions

1. Following are commonly used drugs or biologically important molecules. Classify each as primary, secondary, or tertiary amines or amine salts.

Cocaine
an illicit drug and local anesthetic

Morphine
an opiod analgesic

$C_6H_5-CH_2-CH(NH_2)-COOH$

Phenylalanine
an amino acid and precursor to many neurotransmitters

$CH_2CH_2NH_2$

Serotonin
a neurotransmitter

Methylphenidate (Ritalin)
a stimulant used to treat hyperactivity

$C_6H_5-CH_2-CH(CH_3)-NH_2$

Amphetamine
a stimulant

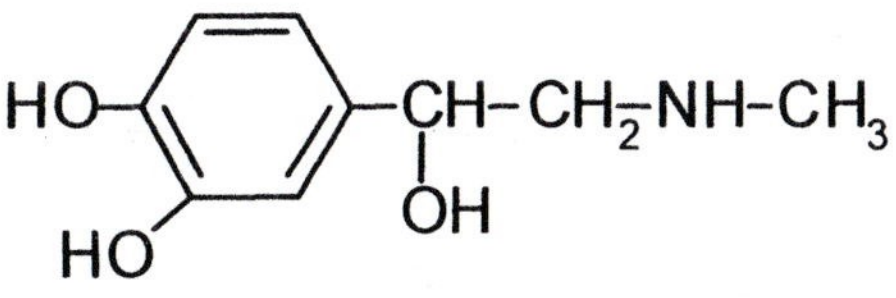

HO, HO, CH–CH$_2$–NH–CH$_3$, OH

Epinephrine (Adrenaline)
a hormone

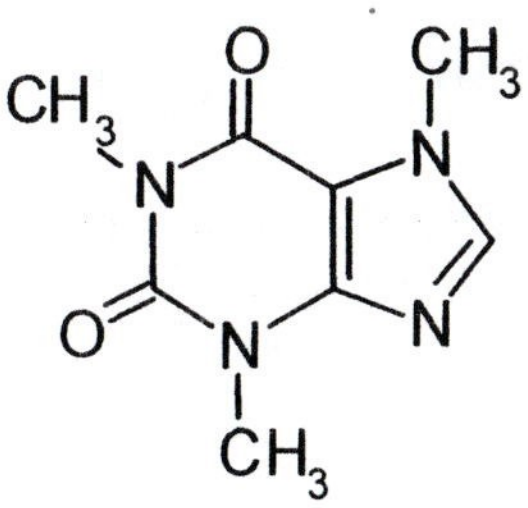

CH_3, O, CH_3, N, N, N, N, O, CH_3

Caffeine

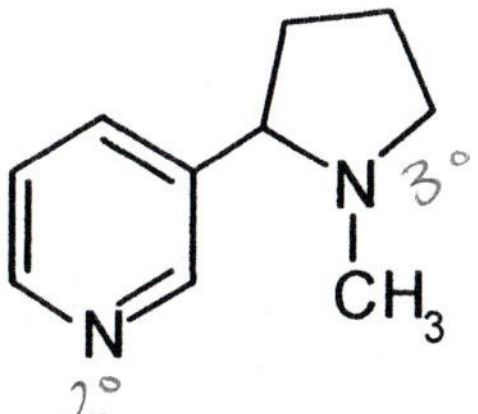

N, CH_3, N, 3°, 2°

Nicotine

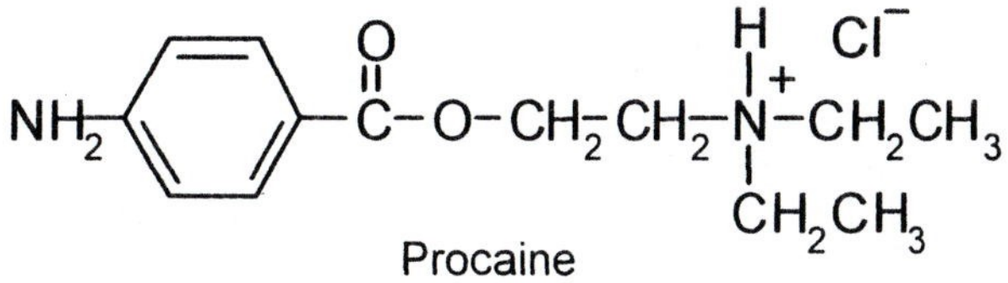

NH_2–C(=O)–O–CH$_2$–CH$_2$–N$^+$(H)–CH$_2$CH$_3$, CH$_2$CH$_3$, Cl$^-$

Procaine
(Novocain)

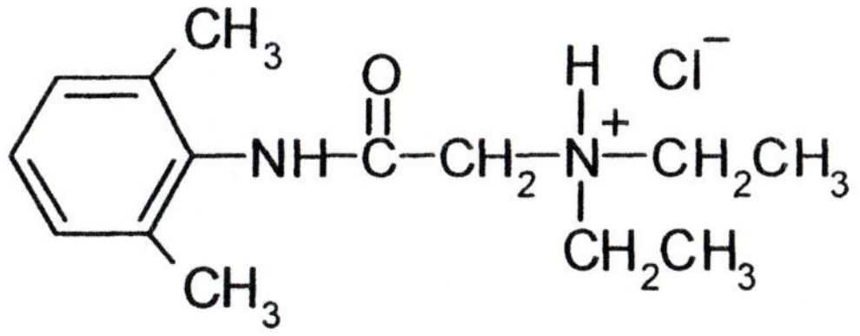

CH_3, CH_3, NH–C(=O)–CH$_2$–N$^+$(H)–CH$_2$CH$_3$, CH$_2$CH$_3$, Cl$^-$

Lidocaine
(Xylocaine)

CH_3, N, O, Cl, N

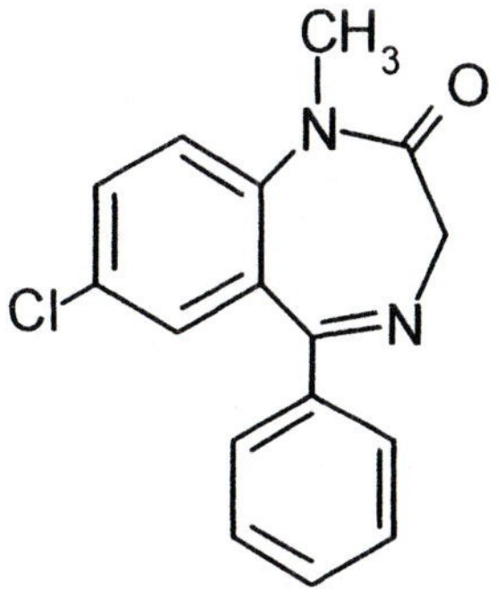

Diazepam
(Valium)

2. Show all the different hydrogen bonds that can be formed in each of the following:

   a. Methylamine

   b. An aqueous solution of methylamine

   c. Acetamide (ethanamide)

   d. An aqueous solution of acetamide (ethanamide)

3. Classify each of the following as acid, base, or neither.

a. CH_3CH_2COONa b. CH_3CH_2OH c. $CH_3CH_2CH(CH_3)_2$

d. CH_3CH_2SH e. CH_3CH_2COOH f. $CH_3CH_2CONH_2$

g. NH_2CH_2COOH h. $CH_3CH{=}CH_2$

i. $C_6H_5{-}\ddot{N}(H){-}C({=}O){-}CH_3$

j. $C_6H_5{-}\ddot{N}(CH_3){-}CH_3$

k. $CH_3CH_2{-}\overset{+}{N}H_3\ Cl^-$

l. $CH_3CH_2{-}\overset{O}{\overset{\|}{C}}{-}OCH_3$

m. $CH_3{-}O{-}\overset{O}{\overset{\|}{P}}(OH){-}OH$

4. Complete the following equations.

a. $CH_3CH_2NH_2 + H_2SO_4 \longrightarrow$?

b. (piperazine: six-membered ring with two N–H groups) + $2\ CH_3COOH \longrightarrow$?

c. ? + ? $\xrightarrow{\text{heat}}$ $H_3C\text{-}\overset{O}{\overset{\|}{C}}\text{-}NH_2 + H_2O$

d. $CH_3CH_2CH_2NH_3^+Br^- + NaOH \xrightarrow{\text{heat}}$? + ? + ?

e. $C_6H_5\text{-}\overset{O}{\overset{\|}{C}}\text{-}N(CH_3)_2 + NaOH \xrightarrow{\text{heat}}$? + ?

f. ? + ? $\longrightarrow CH_3CH_2\text{-}\overset{O}{\overset{\|}{C}}\text{-}OH + NH_3^+Cl^-$

5. Examine the structures of morphine and cocaine (Question 1). These compounds are not water soluble. For parenteral (intravenous or subcutaneous) administration they need to be water soluble. How could these drugs be made more water soluble? Explain with words and structures.

6. The physiologic effects of the following analgesics are similar to those of morphine. Examine their structures and develop an explanation for why this is so.

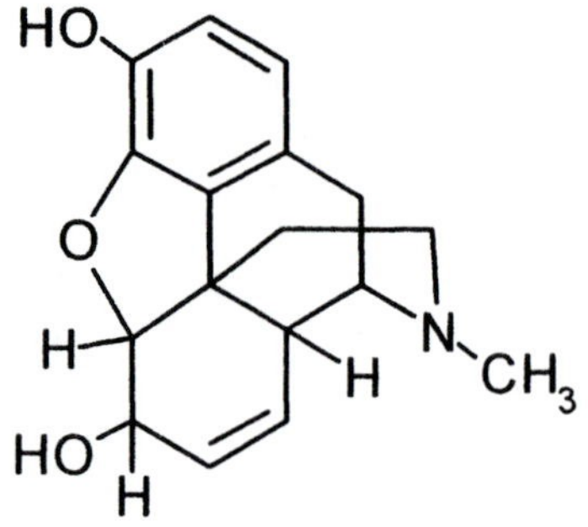

Morphine
an opioid analgesic

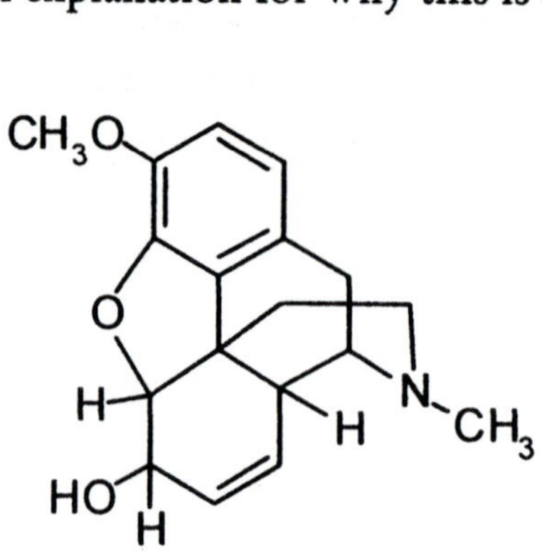

Codeine

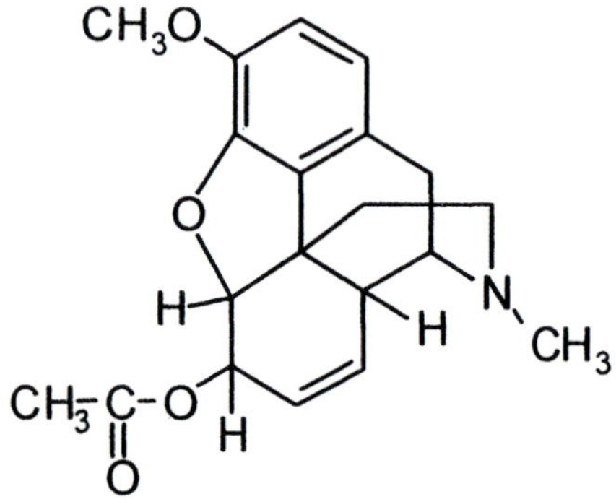

Heroin

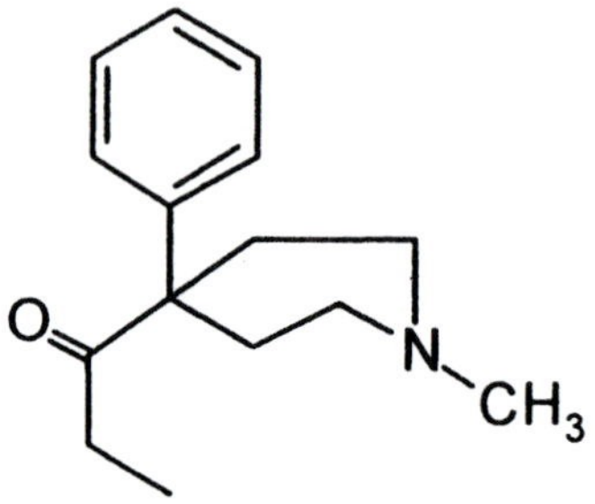

Meperidine
(Demerol)

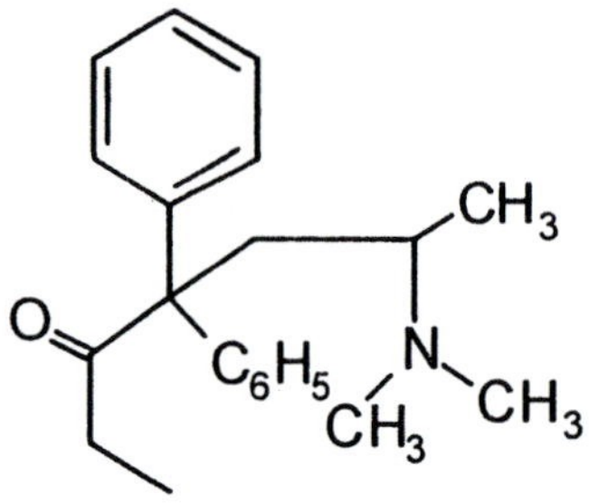

Methadone

7. Reconsider your original answers in the *Topics to Review* section. How has your understanding of these topics changed as a result of this Workshop? Summarize your modified answers in the space below.

Stereochemistry

Topics to Review

Models will be needed for this Workshop.
Review the following tree diagram before answering the questions that follow.

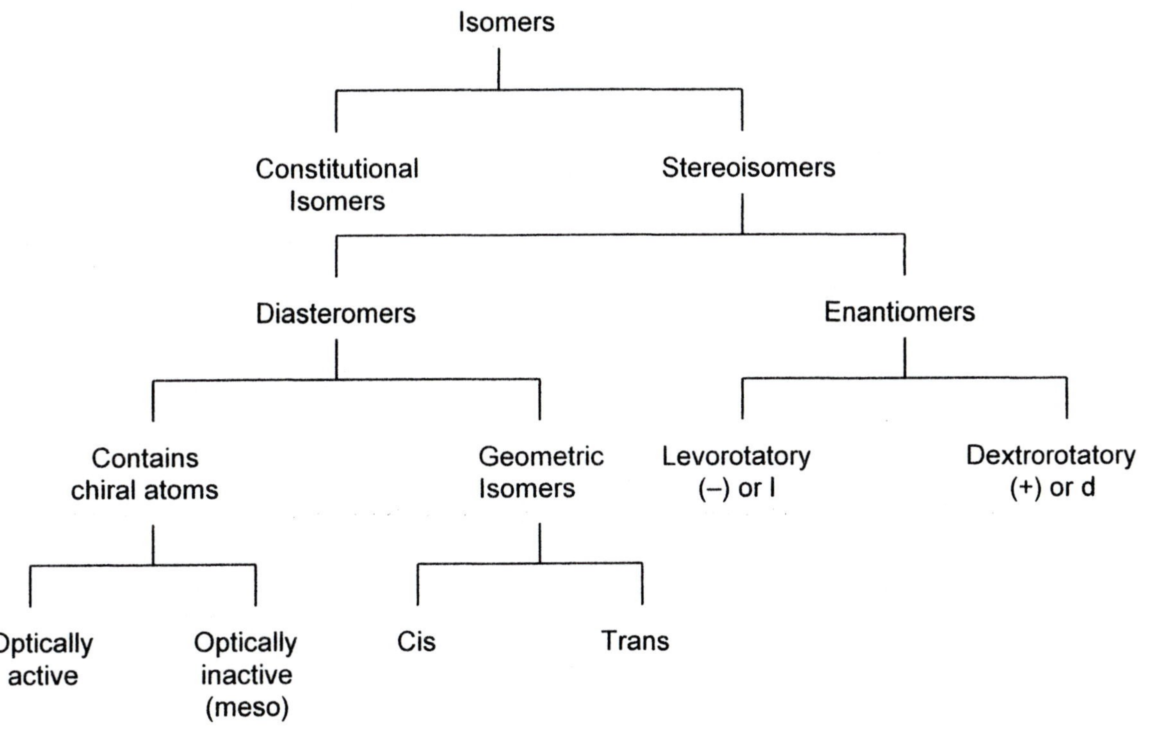

1. Define each of the following terms and in each case write structures of compounds that illustrate the definition.

   a. Isomers

      Textbook pages ___ *to* ___.

      Definition

      Example

   b. Constitutional isomers

      Textbook pages ___ *to* ___.

      Definition

      Example

   c. Stereoisomers

      Textbook pages ___ *to* ___.

      Definition

      Example

d. Diastereomers

Textbook pages ___ *to* ___.

Definition

Example

e. Geometric isomers

Textbook pages ___ *to* ___.

Definition

Example

f. Enantiomers

Textbook pages ___ *to* ___.

Definition

Example

g. Levorotatory

Textbook pages ___ *to* ___.

Definition

Example

h. Dextrorotatory

Textbook pages ___ *to* ___.

Definition

Example

Questions

1. Which of the following molecules are chiral? Identify all chiral carbons in the molecules that are chiral.

a. $H_3C{-}C(CH_2CH_3)(Br){-}CH_2CH_2CH_3$

b. $HOCH_2{-}C({=}O){-}CH_2{-}CH_2OH$ (drawn as CH_2OH / $C{=}O$ / $H{-}C{-}H$ / CH_2OH)

c. cyclohexane–OH

d. $H{-}C(NH_2)(H){-}CO_2H$

e. cyclohexane–CH_2Br

f. $CH_3CH_2{-}CHCH_3$ with Br on the CH

g. cyclohexane with OH and Br (1,3-substituted)

h. cyclohexane with OH and two Br (1,3,5-substituted)

2. Draw all the chiral molecules you can using the molecular formula $C_5H_9Cl_3$.

3. a. Draw a chiral noncyclic hydrocarbon containing six carbons.

   b. Draw its enantiomer.

4. The IUPAC name for glucose is 2,3,4,5,6-pentahydroxyhexanal.

   a. Draw 2,3,4,5,6-pentahydroxyhexanal.

   b. How many chiral carbons does this structure have?

   c. How many stereoisomers would this structure have? Explain.

   d. How many pairs of enantiomers are possible for this structure?

   e. Using the Fischer projection draw at least two pairs of enantiomers.

   f. Will there be any meso stereoisomers? Why or why not?

g. Draw the structure of the hemiacetal that will form when the hydroxyl group on carbon five reacts with the aldehyde group.

h. Does the number of chiral carbons remain the same when the hemiacetal is formed?

5. Thalidomide was an antiemetic (prevents vomiting) given to pregnant women with morning sickness in the 1950s. It was removed from the market when it was discovered to cause phocomelia (absence of limbs) in newborns. It was later learned that only the (-) isomer caused the problem. However, when pure (+) thalidomide was given to animals it was found to racemize. Thus, administration of the pure (+) enantiomer would not necessarily eliminate thalidomide's teratogenicity (causing birth defects).

a. Circle the chiral carbon in the structure of thalidomide shown here.

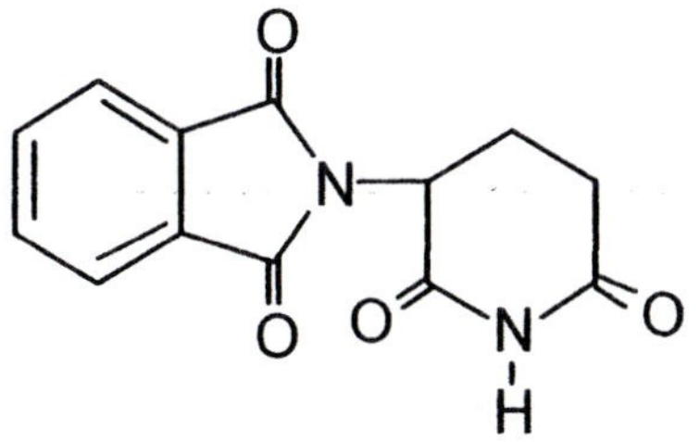

b. Divide the Workshop group into two groups. Each group should write a definition of racemization, and the two groups should compare their definitions and edit them to come up with one definition.

c. How can racemization be detected?

d. Thalidomide racemizes quite easily. What change needs to take place for this to occur?

6. Reconsider your original answers in the *Topics to Review* section. How has your understanding of these topics changed as a result of this Workshop? Summarize your modified answers in the space below.

Carbohydrates

Topics to Review

1. Define the following.

   a. Monosaccharide

   Textbook pages ___ *to* ___.

   b. Disaccharide

   Textbook pages ___ *to* ___.

   c. Polysaccharide

   Textbook pages ___ *to* ___.

   d. Glycosidic linkage

   Textbook pages ___ *to* ___.

e. Anomers

Textbook pages ___ *to* ___.

f. Mutarotation

Textbook pages ___ *to* ___.

g. Pyranose

Textbook pages ___ *to* ___.

h. Furanose

Textbook pages ___ *to* ___.

2. Write the general reaction for the formation of a glycoside from a monosaccharide.

Textbook pages ___ *to* ___.

3. Write the structures for the following:

a. d-glucose

Textbook pages ___ to ___.

b. d-galactose

Textbook pages ___ to ___.

c. d-fructose

Textbook pages ___ to ___.

Questions

1. Consider the structure of d-glucose.

   a. How many anomers does glucose have?

   b. What are their names?

   c. How are they formed? Show with the help of structures.

   d. Are the two anomers of glucose enantiomers? Explain your answer in terms of their optical rotation and structures.

2. Normally, aldehydes are easily oxidized, but ketones are not. However, aldoses and ketoses are both easily oxidized.

   a. Why are ketoses easily oxidized but ketones are not?

   b. Why are carbohydrates that are easily oxidized called *reducing sugars*?

   c. What determines if a carbohydrate is reducing?

   d. Why is this information important?

3. Consider the following two structures. One is glucose and the other is a glucoside.

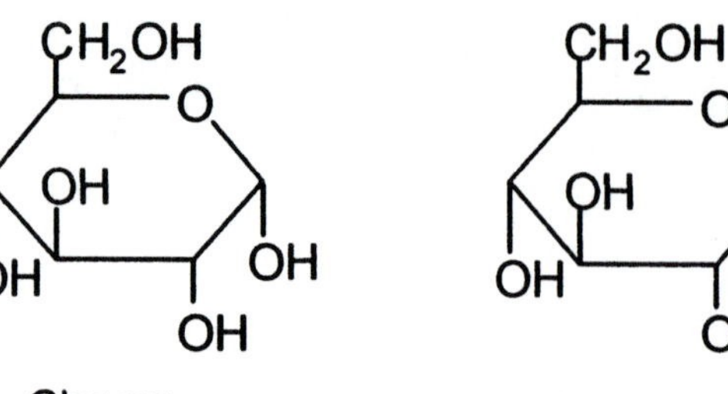

a. Give the reagents and conditions that would change glucose to the glucoside.

b. Name a simple color test that you could use to distinguish between these two compounds. What will be observed in each case?

c. In glucose, what gives rise to mutorotation?

d. Will the glucoside exhibit mutarotation? Explain your answer clearly.

4. Complete the following table.

Disaccharide	Monomer(s)	Type of Glycosidic Linkage	Reducing or Nonreducing
maltose			
lactose			
sucrose			

5. Both glucose and fructose are hexoses, yet when they cyclize, glucose forms a six membered ring (pyranose) whereas fructose forms a five membered ring (furanose). Explain with the help of structures. Why is there a difference in the sizes of the rings formed?

6. Complete the following table.

Polymer	Monomer	Type of Glycosidic Linkage	Linear or Branched	Biological Function
amylose				
amylopectic				
cellulose				
glycogen				

7. A solution of iodine can be used to distinguish between starch and cellulose. Explain with the help of structures. What is the difference between starch and cellulose that makes this possible?

8. Reconsider your original answers in the *Topics to Review* section. How has your understanding of these topics changed as a result of this Workshop? Summarize your modified answers in the space below.

Lipids

Topics to Review

1. What are the characteristics of naturally occurring fatty acids?

   Textbook pages ___ *to* ___.

2. What is the general structure of wax?

   Textbook pages ___ *to* ___.

3. What is the general structure of triacylglycerols (triglycerides)?

Textbook pages ___ *to* ___.

4. Which reactions do triacylglycerols (triglycerides) undergo?

Textbook pages ___ *to* ___.

5. What is the general structure of soap?

Textbook pages ___ *to* ___.

6. What does the term *amphipathic* mean?

Textbook pages ___ *to* ___.

7. What are phospholipids?

Textbook pages ___ *to* ___.

8. Write the general structure of the following:

Textbook pages ___ *to* ___.

a. Glycerophospholipid

b. Sphingomyelin

9. What are glycolipids?

Textbook pages ___ *to* ___.

10. What is the structure of cholesterol? Why is it called a nonsaponifiable lipid?

Textbook pages ___ *to* ___.

11. What is the nature of a cell membrane?

Textbook pages ___ *to* ___.

Questions

1. Lipids are a diverse group of molecules of biological origin that dissolve in nonpolar solvents. Since these molecules are not grouped together on the basis of their structures, it is difficult to remember the structures and functions of lipids. As a group, make a map that will help you remember the different classes of lipids. Assign a function to each class of lipid that you have listed.

2. Consider the structure of fatty acids.

   a. What is the configuration (*cis* or *trans*) of double bonds in most naturally occurring fatty acids?

   b. Generally, the higher the degree of unsaturation in a fatty acid, the lower its melting point. Explain why this is the case.

   c. Which fatty acids are linked to a higher risk of heart disease and cancer?

3. The following triglyceride can undergo several chemical reactions:

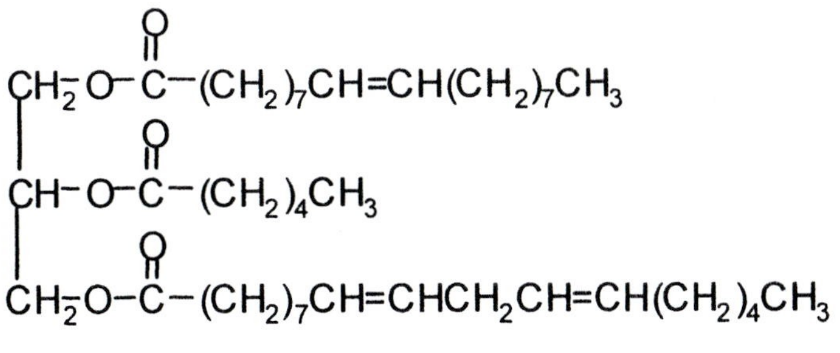

$$\begin{array}{l} CH_2-O-\overset{O}{\overset{\|}{C}}-(CH_2)_7CH=CH(CH_2)_7CH_3 \\ | \\ CH-O-\overset{O}{\overset{\|}{C}}-(CH_2)_4CH_3 \\ | \\ CH_2-O-\overset{O}{\overset{\|}{C}}-(CH_2)_7CH=CHCH_2CH=CH(CH_2)_4CH_3 \end{array}$$

For each of the following reactions list the reactants and reaction conditions needed for the reaction to occur. Explain to the group which part of the molecule will react and what the products will be.

a. Hydrolysis (a reaction that occurs when a fat or oil becomes rancid)

Reactants

Reaction conditions

Products

b. Saponification (a reaction used to produce soap)

Reactants

Reaction conditions

Products

c. Partial hydrogenation (a reaction used to produce margarine from vegetable oil)

Reactants

Reaction conditions

Products

d. Complete hydrogenation (a reaction used to produce solid shortening from vegetable oil)

Reactants

Reaction conditions

Products

e. Oxidation (a reaction that occurs when a lipid becomes rancid)

Reactants

Reaction conditions

Products

4. Lipids are responsible for several features of biological membranes.

   a. Explain how a high concentration of unsaturated hydrocarbon chains makes the membrane more fluid.

   b. Explain why the presence of cholesterol decreases the fluidity of cell membranes.

   c. Explain why lipids make a cell membrane impenetrable to ionic and polar substances.

   d. When lipid bilayers are disrupted they are able to spontaneously reseal. Why?

5. Both of the following reactions are called *saponification*. Does that mean that the product in each case is soap? Explain.

$$CH_3CH_2-\overset{O}{\overset{\|}{C}}-OCH_3 + NaOH \longrightarrow CH_3CH_2-\overset{O}{\overset{\|}{C}}-O^- Na^+ + CH_3OH$$

$$CH_3(CH_2)_{16}-\overset{O}{\overset{\|}{C}}-OCH_3 + NaOH \longrightarrow CH_3(CH_2)_{16}-\overset{O}{\overset{\|}{C}}-O^- Na^+ + CH_3OH$$

6. When grease and soap are mixed an emulsion forms.

   a. Draw a picture to illustrate this.

   b. Soap forms scum (insoluble salts) in hard water. Which ions are present in water that cause the formation of scum?

   c. Design some molecules that would act like soap but could not form insoluble salts in hard water.

7. What is the difference between a micelle, a lipid bilayer vesicle, and a cell membrane? Use diagrams to illustrate the differences.

   a. Discuss the fluid-mosaic model of membranes.

   b. Why is the membrane described as fluid?

   c. Why is it a mosaic?

   d. What is the function of proteins in the cell membrane.

   e. Explain why the percentages of protein, lipid, and carbohydrate vary depending on the type of membrane.

8. Reconsider your original answers in the *Topics to Review* section. How has your understanding of these topics changed as a result of this Workshop? Summarize your modified answers in the space below.

Amino Acids, Peptides, and Proteins

Topics to Review

1. What is the general structure of an l-α-amino acid?

   Textbook pages ___ *to* ___.

2. What are the three-letter and one-letter abbreviations of the commonly occurring α–amino acids?

   Textbook pages ___ *to* ___.

3. What is a zwitterion?

Textbook pages ___ *to* ___.

4. Amino acids can act as acids as well as bases. Write the general equation representing the reaction of an amino acid with a base such as NaOH.

Textbook pages ___ *to* ___.

5. Write the general equation representing the reaction of an amino acid with an acid such as HCl.

Textbook pages ___ *to* ___.

6. What is an isoelectric point?

Textbook pages ___ *to* ___.

7. What is electrophoresis? What is this technique used for?

Textbook pages ___ *to* ___.

8. What reaction does the amino acid cysteine undergo that makes it special?

Textbook pages ___ *to* ___.

9. What are the different levels of protein structure?

Textbook pages ___ *to* ___.

10. What is the native structure of proteins? What is meant by the term denaturation of proteins? What causes proteins to denature?

Textbook pages ___ *to* ___.

Questions

1. Identify all the ionizable groups in Asp, Ala, and Arg.

2. Write the structure of Ala at pH 2, at its pI, and at pH 12.

3. Write the structure of Lys at pH 2, at its pI, and at pH 12.

4. Write the structure of Asp at pH 2, at its pI, and at pH 12.

5. A mixture of Ala, Asp and Lys is subjected to electrophoresis at pH 6.0. Which amino acid will remain stationary? Which will migrate toward the positive electrode, and which will migrate toward the negative electrode?

6. Write all possible tripeptides that can be formed from Ala, Gly, and Val. Do not repeat an amino acid. In each case, identify the C-terminal and the N-terminal.

7. What is the difference between Gly-Ala-Val and Val-Ala-Gly?

8. How many isomeric peptides can be formed from five different amino acids without repeating a single amino acid in the sequence?

9. Design a map to outline the four different levels of protein structure. List all possible structures that belong to each level. At each level list the interactions and forces that are involved in stabilizing the protein structure.

10. Consider the globular protein albumin. Discuss the effects of each of the following denaturing agents on this protein. Which covalent and noncovalent interactions will be affected. In each case, state which levels of protein structure are affected. Use drawings to explain the effect of each agent.

a. Heat

b. UV light

c. Acid

d. Base

e. Ethanol

f. Heavy-metal cations such as Pb^{2+} and Hg^{2+}

g. Urea, NH_2CONH_2

11. When a mutation occurs, substitution of an amino acid for one with similar properties is called a *conservative* substitution. In contrast, a *nonconservative* substitution occurs when an amino acid is replaced by one of very different polarity or volume.

    a. Which substitution, conservative or nonconservative is usually the most harmful? Why?

    b. For each of the following give an example of both types of substitution.

Amino Acid	Conservative	Nonconservative
Val		
Cys		
Phe		
Glu		
Arg		

12. Wool is a protein that is rich in the amino acid cysteine. Hat makers of the past used mercury salts as reagents for shaping wool. The term "mad hatter" evolved as mercury, a toxin, had adverse effects on the hat makers who worked with it in large quantities. Consider the chemistry behind this method of wool shaping.

a. Which functional groups of wool help it retain its shape?

b. What reaction occurs when wool reacts with the mercury ion? (*Hint*: Mercaptan means "mercury grabbing.")

c. How does the chemical reaction change the properties of wool?

d. How does the reverse reaction affect the properties of wool?

e. Compare the process of hat making to that of permanent-waving hair.

13. Reconsider your original answers in the *Topics to Review* section. How has your understanding of these topics changed as a result of this Workshop? Summarize your modified answers in the space below.

Enzymes

Topics to Review

1. What are enzymes?

   Textbook pages ___ *to* ___.

2. What are substrates?

   Textbook pages ___ *to* ___.

3. How are enzymes named, and what are the general classes of enzymes?

   Textbook pages ___ to ___.

4. What is an active site?

   Textbook pages ___ to ___.

5. What are the two most popular models used for explaining the mechanism of enzyme action? What are the differences between these two models?

   Textbook pages ___ to ___.

6. Name the factors that affect enzyme activity, and explain how each factor influences the enzyme's activity.

   Textbook pages ___ *to* ___.

7. What are inhibitors? How many general types of inhibitors are there?

   Textbook pages ___ *to* ___.

8. Make a list of the ways in which enzymes are regulated.

Textbook pages ___ *to* ___.

9. How are isoenzymes used for diagnosis in medicine?

Textbook pages ___ *to* ___.

Questions

1. What is the name of the enzyme that catalyzes each of the following reactions?

a. $NH_2CH_2COOH + HOOC{-}CH_2{-}CH_2{-}C(=O){-}C(=O){-}OH \longrightarrow H{-}C(=O){-}COOH + HOOC{-}CH_2{-}CH_2{-}C(H)(NH_2){-}C(=O){-}OH$

b. $CH_3{-}C(=O){-}OCH_2CH_3 \xrightarrow{H_2O} CH_3{-}C(=O){-}OH + CH_3CH_2OH$

c. $^{-}OOC{-}CH(OH){-}CH_2OPO_3^{2-} \rightleftharpoons {}^{-}OOC{-}CH(OPO_3^{2-}){-}CH_2OH$

d. $^{-}OOC{-}CH_2{-}CH_2{-}COO^{-} + FAD \longrightarrow {}^{-}OOC{-}CH{=}CH{-}COO^{-} + FADH_2$

e. $H{-}C(H)(OH){-}C(=O){-}CH_2OPO_3^{2-} \rightleftharpoons H{-}C(=O){-}CH(OH){-}CH_2OPO_3^{2-}$

f. $CH_3{-}C(=O){-}C(=O){-}OH \longrightarrow CH_3{-}C(=O){-}H + CO_2$

2. Draw a graph of reaction rate versus pH for the following:

a. An enzyme active in the mouth

b. An enzyme active in the stomach

c. An enzyme active in the intestines

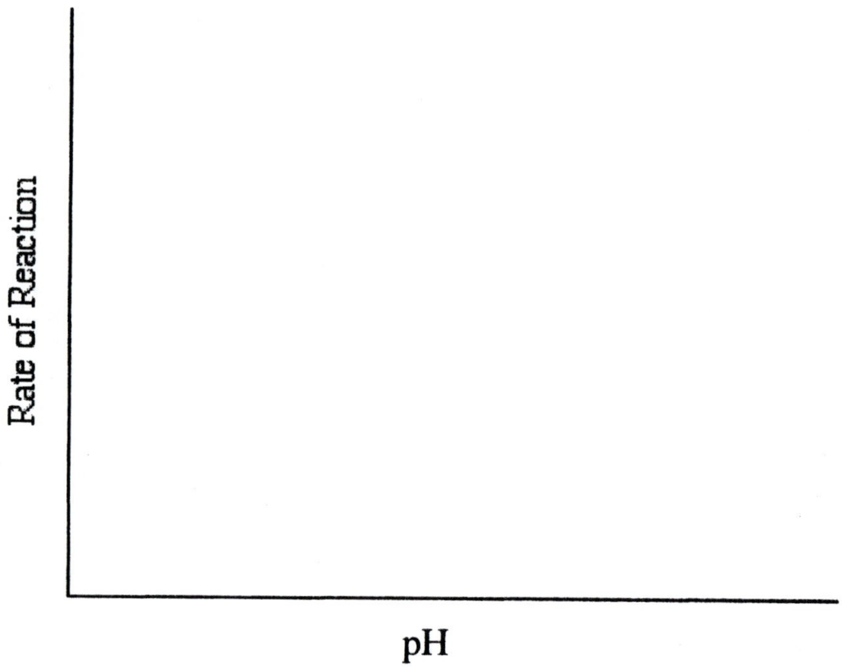

Rate of Reaction

pH

3. Urease catalyzes the hydrolysis of urea. Write the structures of six molecules that might inhibit urease. Each member of the group should try to contribute one structure.

$$NH_2-\overset{\displaystyle O}{\overset{\|}{C}}-NH_2$$

4. The pancreas secretes several digestive enzymes, including α–amylase, proteases, lipases, and nucleases. How does the pancreas protect itself from the destructive effects of these hydrolases?

5. Alcohol dehydrogenase is involved in the oxidation of both methanol, CH_3OH, and ethanol, CH_3CH_2OH. How might ethanol work as an antidote for methanol poisoning?

6. In recent years, certain bacteria have developed a resistance to penicillin due to penicillinase, a bacterial enzyme that cleaves the β-lactam ring of penicillin, rendering it ineffective. However, clavulanic acid, when administered with penicillin, restores the effectiveness of penicillin. Clavulanic acid alone has no antibacterial properties. How does the combination of drugs work?

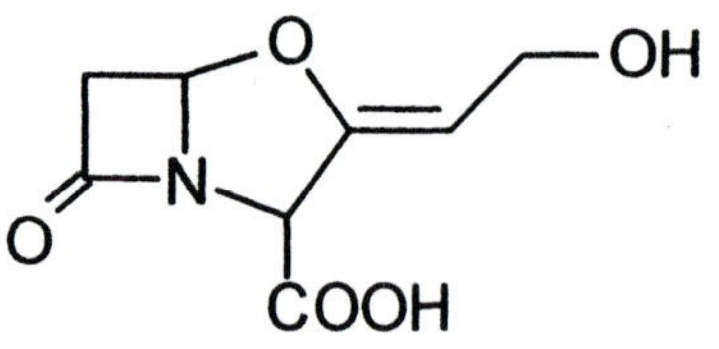

Clavulanic acid

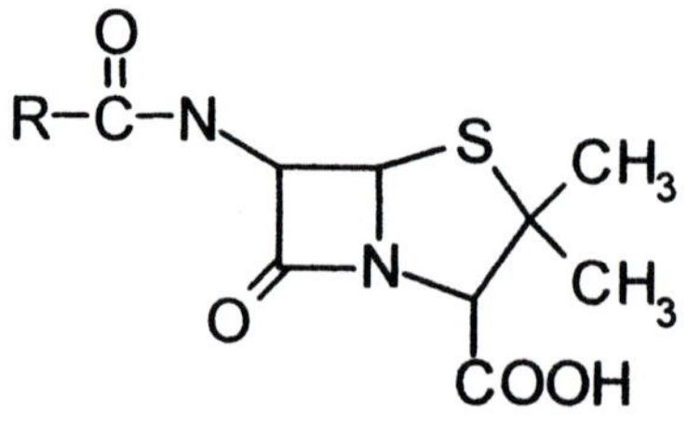

Penicillin

7. In the succinate dehydrogenase reaction, succinate is the substrate.

a. Can you write the structure of the product of this reaction?

```
    O  H  H  O
    ‖  |  |  ‖
⁻O–C–C–C–C–O⁻
       |  |
       H  H
```

succinate

b. Malonate is an inhibitor of this enzyme. What type of an inhibitor is it? Why?

```
    O  H  O
    ‖  |  ‖
⁻O–C–C–C–O⁻
       |
       H
```

malonate

c. In the following figure, which curve would represent the inhibition by malonate?

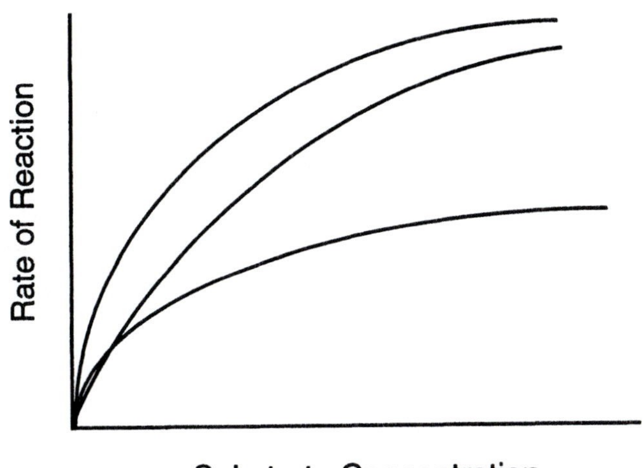

d. Explain why you picked the one you did and then explain what the other two curves represent.

e. Can the effect of malonate be reversed by addition of additional amounts of succinate to the reaction mixture?

f. Draw at least two more molecules that could be inhibitors of succinate dehydrogenase.

8. Explain how enzymes can be used for the following:

   a. Diagnosis of myocardial infarction

   b. Diagnosis of liver disease

   c. Diagnosis of diabetes mellitus

   d. Meat tenderizer

   e. Contact lens cleaner

   f. Cheese manufacturing

   g. What other applications can you think of?

9. Reconsider your original answers in the *Topics to Review* section. How has your understanding of these topics changed as a result of this Workshop? Summarize your modified answers in the space below.

Nucleic Acids and Protein Synthesis

Topics to Review

1. What are the differences among nucleosides, nucleotides, and nucleic acids?

   Textbook pages ___ *to* ___.

2. What is the structure of DNA? Describe both primary and secondary structures.

   Textbook pages ___ *to* ___.

3. What are the differences between DNA and RNA?

   Textbook pages ___ to ___.

4. How is DNA replicated? Outline all the steps.

   Textbook pages ___ to ___.

5. What is transcription?

   Textbook pages ___ to ___.

6. What is translation?

Textbook pages ___ *to* ___.

7. What are the characteristics of the genetic code?

Textbook pages ___ *to* ___.

8. What is mutation?

Textbook pages ___ *to* ___.

9. What is recombinant DNA? How is it made?

Textbook pages ___ *to* ___.

Questions

1. Draw a logic tree to illustrate the process of protein synthesis. Begin with DNA and examine all intermediate stages that contribute to the formation of the polypeptide chain.

2. Consider the following peptide:

Ala-Tyr-Gly-Phe-Phe-Trp-Arg

a. Give a sequence on the mRNA that could code for this peptide. Label the 3′ and 5′ ends.

b. What would be the sequence on the DNA from which this mRNA would be transcribed? Label the 3′ and 5′ ends.

c. What anticodon sequence would be complementary to the codons listed in Question 2a? Label the 3′ and 5′ ends.

3. Consider the following mrNA fragment:

1 3 6 9 12 15 18 21 24 27 30 33 36 39 42
AAUGAUAGGCCUCCGGCACAUCACUUAAGUAGGGAAGUAUAA

State what effect, if any, the following changes will have on the resulting polypeptide:

a. Replacement of uracil in position 6 by guanine

b. Replacement of adenine at position 18 by cytosine

c. Replacement of uracil at position 26 by adenine

d. Deletion of guanine at position 15

e. Which of these mutations is likely to be the most serious? Why?

4. a. Explain how bacteria can be turned into mini drug factories.

b. What is the advantage of using bacterial cells instead of cells from higher organisms?

c. Name some drugs made this way.

d. What is the advantage of using proteins synthesized using recombinant DNA technology over proteins isolated from animal sources.

5. The goal of the Human Genome Project is to determine the total sequence of the genome ($\approx$ 100,000 genes) and to identify the sequence of those genes. In addition, it is hoped that genes governing many diseases will be identified.

   a. What are the advantages of having this kind of information about each person?

   b. What are the problems associated with having this kind of information available?

6. Sickle cell anemia is a genetic disorder most prevalent among people of African and Asian origin. This disease is caused by a single-point mutation that results in the replacement of glutamic acid by valine at the sixth position of the hemoglobin β-chain.

   a. Use the genetic code to determine which bases are involved in this mutation.

   b. How does this single substitution affect the structure and function of hemoglobin?

   c. Discuss some symptoms a patient with sickle cell anemia might experience as well as treatment options that exist now and possible treatments in the future.

   d. What aspect of this mutation can be considered to be an advantage?

7. a. What is DNA fingerprinting?

b. What is it used for?

c. What are its advantages?

d. What are its limitations?

8. a. Explain how sequences of amino acids in different proteins can be used to study the process of evolution.

b. Can mRNA be used for the same purpose?

9. a. Do you think that recombinant DNA technology could be used to "meddle with evolution"? How?

b. Give specific examples and discuss the pertinent ethical issues in each case. Each group should provide at least four examples.

10. Reconsider your original answers in the *Topics to Review* section. How has your understanding of these topics changed as a result of this Workshop? Summarize your modified answers in the space below.

Carbohydrate Metabolism

Topics to Review

1. What is the difference between digestion and metabolism?

   Textbook pages ___ *to* ___.

2. What is catabolism?

   Textbook pages ___ *to* ___.

3. What is anabolism?

Textbook pages ___ to ___.

4. List at least three differences between catabolic reactions and anabolic reactions.

Textbook pages ___ to ___.

5. What is the structure of ATP? What is the difference between the structures of ATP, ADP, and AMP?

Textbook pages ___ to ___.

6. Write equations to show the conversion of ATP into ADP and AMP.

Textbook pages ___ *to* ___.

7. What does NAD^+ stand for? Which vitamin is essential for the synthesis of NAD^+? What is the function of NAD^+?

Textbook pages ___ *to* ___.

8. What does FAD stand for? Which vitamin is essential for the synthesis of FAD? What is the function of FAD?

Textbook pages ___ *to* ___.

9. What is the role of coenzyme A? Which vitamin is essential for the synthesis of coenzyme A?

   Textbook pages ___ *to* ___.

10. What is glycolysis?

    Textbook pages ___ *to* ___.

11. What is the purpose of the citric acid cycle?

    Textbook pages ___ *to* ___.

12. What is oxidative phosphorylation?

Textbook pages ___ *to* ___.

Questions

1. Monosaccharides are degraded via glycolysis.

   a. Where in the cell does glycolysis occur?

   b. Write the overall reaction for glycolysis.

   c. Is the rate of glycolysis affected by the availability of oxygen?

   d. Is sufficient oxygen available to the muscles during periods of high activity?

   e. Write a reaction to show what happens to pyruvate (end product of glycolysis) in the muscles during periods of high activity.

f. Is pyruvate oxidized or reduced under these conditions? What is the oxidizing/reducing agent? Is this reaction reversible?

g. Write an equation to show what happens to pyruvate in the presence of oxygen. Is this reaction reversible?

2. The TCA cycle is the abbreviation for the tricarboxylic acid cycle.

   a. Where in the cell does the TCA cycle occur?

   b. Why is this set of reactions called a cycle?

   c. Write the overall reaction for the TCA cycle.

   d. What is the waste product of the TCA cycle?

   e. Which product(s) of the TCA cycle is(are) oxidized in the electron transport chain?

3. ETC is the abbreviation for the electron transport chain.

a. Where in the cell do the reactions that constitute the ETC occur?

b. Write the overall reaction for the ETC.

c. What is the waste product of the ETC?

d. Where does the oxygen required for the oxidation reactions in ETC come from?

e. What happens if the oxygen supply is cut off?

f. Explain the difference between oxidative phosphorylation and substrate phosphorylation. Give examples of each.

g. Explain the effect of ingestion of cyanide ion on the ETC.

4. The TCA cycle operates to meet the cellular needs for ATP. When levels of ATP are high the rate at which the cycle operates is reduced. Where in the cycle is the control most likely to occur?

5. Divide the Workshop group into two smaller groups. One group should calculate the number of moles of ATP produced when one mole of glucose is catabolized in the liver. The other group should calculate the number of moles of ATP produced when one mole of glucose is catabolized in the muscle. Select a spokesperson from each group to present the results to the whole group. Are the results of the two calculations the same? Should they be the same? Discuss why or why not.

6. Reconsider your original answers in the *Topics to Review* section. How has your understanding of these topics changed as a result of this Workshop? Summarize your modified answers in the space below.

Lipid and Protein Metabolism

Topics to Review

1. Name the pathway used in the cells by which energy is extracted from glycerol. Where in the cells does this pathway occur?

   Textbook pages ___ *to* ___.

2. Name the pathway used by the cells to extract energy from fatty acids. Where in the cells are the enzymes located to catalyze the reactions in this pathway?

   Textbook pages ___ *to* ___.

3. Name the compound into which the fatty acids are broken down before they enter the TCA cycle.

   Textbook pages ___ *to* ___.

4. Write the structures of the three compounds referred to as ketone bodies. How many of these are actually ketones?

   Textbook pages ___ *to* ___.

5. When proteins are digested most of the α-amino acids produced cross the intestinal wall and enter the amino acid pool to serve as building blocks for proteins. Under what conditions are amino acids used as fuel for energy?

   Textbook pages ___ *to* ___.

6. Write the general reaction for oxidative deamination.

Textbook pages ___ *to* ___.

7. What is the urea cycle? Where in the cells are the enzymes that catalyze the reactions in this cycle located?

Textbook pages ___ *to* ___.

Questions

1. Consider the metabolism of the following triacylglycerol (triglyceride).

$$
\begin{array}{l}
H_2C-O-\overset{\displaystyle O}{\overset{\|}{C}}-(CH_2)_{12}CH_3 \\
\;|\\
HC-O-\overset{\displaystyle O}{\overset{\|}{C}}-(CH_2)_{14}CH_3 \\
\;|\\
H_2C-O-\overset{\displaystyle O}{\overset{\|}{C}}-(CH_2)_{16}CH_3
\end{array}
$$

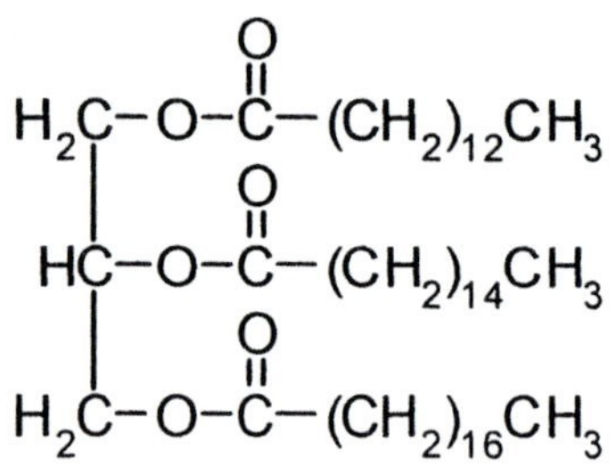

a. What happens to the glycerol that is generated upon hydrolysis of this triacylglycerol (triglyceride)?

b. How many moles of ATP are generated per mole of glycerol?

c. Break the group up into pairs. Have each pair calculate the number of moles of ATP generated from one of the fatty acids produced on hydrolysis of one mole of the triacylglycerol (triglyceride) at the beginning of this question. The pairs should report the results to the whole group.

d. The whole group should calculate the total number of moles of ATP generated by the catabolism of one mole of the entire triacylglycerol.

2. Discuss all possible fates of acetyl CoA generated by β–oxidation of fatty acids.

3. Oxaloacetate is required for the acetyl CoA generated during β–oxidation of fatty acids to enter the TCA cycle. Low glucose supply slows down the production of oxaloacetate.

   a. What is the connection between glucose supply and the production of oxaloacetate?

   b. Name three pathological conditions under which glucose supply is low.

   c. Low glucose supply causes a buildup of acetyl CoA which in turn promotes ketogenesis. In which organ are ketone bodies produced?

   d. What happens to blood pH when excess ketone bodies are produced? Explain.

   e. What is this condition called?

   f. How is this condition treated?

4. The reaction that produces ammonium ions from α–amino acids is called oxidative deamination.

   a. Write the general reaction for oxidative deamination.

   b. Identify the molecule that is oxidized in this reaction.

   c. What is the oxidizing agent?

5. The ammonium ions produced by oxidative deamination are converted to a less toxic substance in the liver.

   a. Name the substance to which the ammonium ions are converted before they are excreted in the urine.

   b. What is the metabolic cycle called that detoxifies the ammonium ions?

   c. Write the overall reaction for this cycle.

   d. Is this reaction endergonic or exergonic?

6. Explain how genetic defects give rise to metabolic diseases. In each of the following explain the genetic defect that gives rise to the disease.

   a. Hyperammonemia

   b. PKU

   c. Cori's disease (glycogen storage disease)

   d. Tay-Sachs disease (lipid storage disease)

7. As a group draw a diagram that connects the catabolic pathways of carbohydrates, lipids, and proteins. The team leader should take the role of a scribe while the other team members provide the connections.

8. Reconsider your original answers in the *Topics to Review* section. How has your understanding of these topics changed as a result of this Workshop? Summarize your modified answers in the space below.